DESIGN OF
NUCLEAR INSTALLATIONS
AGAINST EXTERNAL EVENTS
EXCLUDING EARTHQUAKES

The following States are Members of the International Atomic Energy Agency:

AFGHANISTAN	GEORGIA	OMAN
ALBANIA	GERMANY	PAKISTAN
ALGERIA	GHANA	PALAU
ANGOLA	GREECE	PANAMA
ANTIGUA AND BARBUDA	GRENADA	PAPUA NEW GUINEA
ARGENTINA	GUATEMALA	PARAGUAY
ARMENIA	GUYANA	PERU
AUSTRALIA	HAITI	PHILIPPINES
AUSTRIA	HOLY SEE	POLAND
AZERBAIJAN	HONDURAS	PORTUGAL
BAHAMAS	HUNGARY	QATAR
BAHRAIN	ICELAND	REPUBLIC OF MOLDOVA
BANGLADESH	INDIA	ROMANIA
BARBADOS	INDONESIA	RUSSIAN FEDERATION
BELARUS	IRAN, ISLAMIC REPUBLIC OF	RWANDA
BELGIUM	IRAQ	SAINT LUCIA
BELIZE	IRELAND	SAINT VINCENT AND
BENIN	ISRAEL	THE GRENADINES
BOLIVIA, PLURINATIONAL	ITALY	SAMOA
STATE OF	JAMAICA	SAN MARINO
BOSNIA AND HERZEGOVINA	JAPAN	SAUDI ARABIA
BOTSWANA	JORDAN	SENEGAL
BRAZIL	KAZAKHSTAN	SERBIA
BRUNEI DARUSSALAM	KENYA	SEYCHELLES
BULGARIA	KOREA, REPUBLIC OF	SIERRA LEONE
BURKINA FASO	KUWAIT	SINGAPORE
BURUNDI	KYRGYZSTAN	SLOVAKIA
CAMBODIA	LAO PEOPLE'S DEMOCRATIC	SLOVENIA
CAMEROON	REPUBLIC	SOUTH AFRICA
CANADA	LATVIA	SPAIN
CENTRAL AFRICAN	LEBANON	SRI LANKA
REPUBLIC	LESOTHO	SUDAN
CHAD	LIBERIA	SWEDEN
CHILE	LIBYA	SWITZERLAND
CHINA	LIECHTENSTEIN	SYRIAN ARAB REPUBLIC
COLOMBIA	LITHUANIA	TAJIKISTAN
COMOROS	LUXEMBOURG	THAILAND
CONGO	MADAGASCAR	TOGO
COSTA RICA	MALAWI	TRINIDAD AND TOBAGO
CÔTE D'IVOIRE	MALAYSIA	TUNISIA
CROATIA	MALI	TURKEY
CUBA	MALTA	TURKMENISTAN
CYPRUS	MARSHALL ISLANDS	UGANDA
CZECH REPUBLIC	MAURITANIA	UKRAINE
DEMOCRATIC REPUBLIC	MAURITIUS	UNITED ARAB EMIRATES
OF THE CONGO	MEXICO	UNITED KINGDOM OF
DENMARK	MONACO	GREAT BRITAIN AND
DJIBOUTI	MONGOLIA	NORTHERN IRELAND
DOMINICA	MONTENEGRO	UNITED REPUBLIC
DOMINICAN REPUBLIC	MOROCCO	OF TANZANIA
ECUADOR	MOZAMBIQUE	UNITED STATES OF AMERICA
EGYPT	MYANMAR	URUGUAY
EL SALVADOR	NAMIBIA	UZBEKISTAN
ERITREA	NEPAL	VANUATU
ESTONIA	NETHERLANDS	VENEZUELA, BOLIVARIAN
ESWATINI	NEW ZEALAND	REPUBLIC OF
ETHIOPIA	NICARAGUA	VIET NAM
FIJI	NIGER	YEMEN
FINLAND	NIGERIA	ZAMBIA
FRANCE	NORTH MACEDONIA	ZIMBABWE
GABON	NORWAY	

The Agency's Statute was approved on 23 October 1956 by the Conference on the Statute of the IAEA held at United Nations Headquarters, New York; it entered into force on 29 July 1957. The Headquarters of the Agency are situated in Vienna. Its principal objective is "to accelerate and enlarge the contribution of atomic energy to peace, health and prosperity throughout the world".

IAEA SAFETY STANDARDS SERIES No. SSG-68

DESIGN OF NUCLEAR INSTALLATIONS AGAINST EXTERNAL EVENTS EXCLUDING EARTHQUAKES

SPECIFIC SAFETY GUIDE

INTERNATIONAL ATOMIC ENERGY AGENCY
VIENNA, 2021

COPYRIGHT NOTICE

Marketing and Sales Unit, Publishing Section
International Atomic Energy Agency
Vienna International Centre
PO Box 100
1400 Vienna, Austria
fax: +43 1 26007 22529
tel.: +43 1 2600 22417
email: sales.publications@iaea.org
www.iaea.org/publications

© IAEA, 2021

Printed by the IAEA in Austria
December 2021
STI/PUB/1968

IAEA Library Cataloguing in Publication Data

Names: International Atomic Energy Agency.
Title: Design of nuclear installations against external events excluding earthquakes / International Atomic Energy Agency.
Description: Vienna : International Atomic Energy Agency, 2021. | Series: IAEA safety standards series, ISSN 1020–525X ; no. SSG-68 | Includes bibliographical references.
Identifiers: IAEAL 21-01455 | ISBN 978–92–0–136021–2 (paperback : alk. paper) | ISBN 978–92–0–136121–9 (pdf) | ISBN 978–92–0–136221–6 (epub)
Subjects: LCSH: Nuclear facilities — Design and construction — Safety measures. | Nuclear facilities — Safety measures. | Nuclear power plants — Natural disaster effects.
Classification: UDC 621.039.58 | STI/PUB/1968

FOREWORD

by Rafael Mariano Grossi
Director General

The IAEA's Statute authorizes it to "establish...standards of safety for protection of health and minimization of danger to life and property". These are standards that the IAEA must apply to its own operations, and that States can apply through their national regulations.

The IAEA started its safety standards programme in 1958 and there have been many developments since. As Director General, I am committed to ensuring that the IAEA maintains and improves upon this integrated, comprehensive and consistent set of up to date, user friendly and fit for purpose safety standards of high quality. Their proper application in the use of nuclear science and technology should offer a high level of protection for people and the environment across the world and provide the confidence necessary to allow for the ongoing use of nuclear technology for the benefit of all.

Safety is a national responsibility underpinned by a number of international conventions. The IAEA safety standards form a basis for these legal instruments and serve as a global reference to help parties meet their obligations. While safety standards are not legally binding on Member States, they are widely applied. They have become an indispensable reference point and a common denominator for the vast majority of Member States that have adopted these standards for use in national regulations to enhance safety in nuclear power generation, research reactors and fuel cycle facilities as well as in nuclear applications in medicine, industry, agriculture and research.

The IAEA safety standards are based on the practical experience of its Member States and produced through international consensus. The involvement of the members of the Safety Standards Committees, the Nuclear Security Guidance Committee and the Commission on Safety Standards is particularly important, and I am grateful to all those who contribute their knowledge and expertise to this endeavour.

The IAEA also uses these safety standards when it assists Member States through its review missions and advisory services. This helps Member States in the application of the standards and enables valuable experience and insight to be shared. Feedback from these missions and services, and lessons identified from events and experience in the use and application of the safety standards, are taken into account during their periodic revision.

I believe the IAEA safety standards and their application make an invaluable contribution to ensuring a high level of safety in the use of nuclear technology. I encourage all Member States to promote and apply these standards, and to work with the IAEA to uphold their quality now and in the future.

THE IAEA SAFETY STANDARDS

BACKGROUND

Radioactivity is a natural phenomenon and natural sources of radiation are features of the environment. Radiation and radioactive substances have many beneficial applications, ranging from power generation to uses in medicine, industry and agriculture. The radiation risks to workers and the public and to the environment that may arise from these applications have to be assessed and, if necessary, controlled.

Activities such as the medical uses of radiation, the operation of nuclear installations, the production, transport and use of radioactive material, and the management of radioactive waste must therefore be subject to standards of safety.

Regulating safety is a national responsibility. However, radiation risks may transcend national borders, and international cooperation serves to promote and enhance safety globally by exchanging experience and by improving capabilities to control hazards, to prevent accidents, to respond to emergencies and to mitigate any harmful consequences.

States have an obligation of diligence and duty of care, and are expected to fulfil their national and international undertakings and obligations.

International safety standards provide support for States in meeting their obligations under general principles of international law, such as those relating to environmental protection. International safety standards also promote and assure confidence in safety and facilitate international commerce and trade.

A global nuclear safety regime is in place and is being continuously improved. IAEA safety standards, which support the implementation of binding international instruments and national safety infrastructures, are a cornerstone of this global regime. The IAEA safety standards constitute a useful tool for contracting parties to assess their performance under these international conventions.

THE IAEA SAFETY STANDARDS

The status of the IAEA safety standards derives from the IAEA's Statute, which authorizes the IAEA to establish or adopt, in consultation and, where appropriate, in collaboration with the competent organs of the United Nations and with the specialized agencies concerned, standards of safety for protection of health and minimization of danger to life and property, and to provide for their application.

With a view to ensuring the protection of people and the environment from harmful effects of ionizing radiation, the IAEA safety standards establish fundamental safety principles, requirements and measures to control the radiation exposure of people and the release of radioactive material to the environment, to restrict the likelihood of events that might lead to a loss of control over a nuclear reactor core, nuclear chain reaction, radioactive source or any other source of radiation, and to mitigate the consequences of such events if they were to occur. The standards apply to facilities and activities that give rise to radiation risks, including nuclear installations, the use of radiation and radioactive sources, the transport of radioactive material and the management of radioactive waste.

Safety measures and security measures[1] have in common the aim of protecting human life and health and the environment. Safety measures and security measures must be designed and implemented in an integrated manner so that security measures do not compromise safety and safety measures do not compromise security.

The IAEA safety standards reflect an international consensus on what constitutes a high level of safety for protecting people and the environment from harmful effects of ionizing radiation. They are issued in the IAEA Safety Standards Series, which has three categories (see Fig. 1).

Safety Fundamentals

Safety Fundamentals present the fundamental safety objective and principles of protection and safety, and provide the basis for the safety requirements.

Safety Requirements

An integrated and consistent set of Safety Requirements establishes the requirements that must be met to ensure the protection of people and the environment, both now and in the future. The requirements are governed by the objective and principles of the Safety Fundamentals. If the requirements are not met, measures must be taken to reach or restore the required level of safety. The format and style of the requirements facilitate their use for the establishment, in a harmonized manner, of a national regulatory framework. Requirements, including numbered 'overarching' requirements, are expressed as 'shall' statements. Many requirements are not addressed to a specific party, the implication being that the appropriate parties are responsible for fulfilling them.

Safety Guides

Safety Guides provide recommendations and guidance on how to comply with the safety requirements, indicating an international consensus that it

---

[1]   See also publications issued in the IAEA Nuclear Security Series.

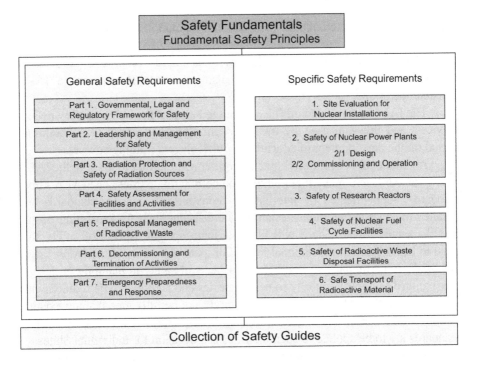

Safety Fundamentals	
Fundamental Safety Principles	
General Safety Requirements	**Specific Safety Requirements**
Part 1. Governmental, Legal and Regulatory Framework for Safety	1. Site Evaluation for Nuclear Installations
Part 2. Leadership and Management for Safety	2. Safety of Nuclear Power Plants 2/1 Design 2/2 Commissioning and Operation
Part 3. Radiation Protection and Safety of Radiation Sources	
Part 4. Safety Assessment for Facilities and Activities	3. Safety of Research Reactors
Part 5. Predisposal Management of Radioactive Waste	4. Safety of Nuclear Fuel Cycle Facilities
Part 6. Decommissioning and Termination of Activities	5. Safety of Radioactive Waste Disposal Facilities
Part 7. Emergency Preparedness and Response	6. Safe Transport of Radioactive Material
Collection of Safety Guides	

FIG. 1. The long term structure of the IAEA Safety Standards Series.

is necessary to take the measures recommended (or equivalent alternative measures). The Safety Guides present international good practices, and increasingly they reflect best practices, to help users striving to achieve high levels of safety. The recommendations provided in Safety Guides are expressed as 'should' statements.

APPLICATION OF THE IAEA SAFETY STANDARDS

The principal users of safety standards in IAEA Member States are regulatory bodies and other relevant national authorities. The IAEA safety standards are also used by co-sponsoring organizations and by many organizations that design, construct and operate nuclear facilities, as well as organizations involved in the use of radiation and radioactive sources.

The IAEA safety standards are applicable, as relevant, throughout the entire lifetime of all facilities and activities — existing and new — utilized for peaceful purposes and to protective actions to reduce existing radiation risks. They can be used by States as a reference for their national regulations in respect of facilities and activities.

The IAEA's Statute makes the safety standards binding on the IAEA in relation to its own operations and also on States in relation to IAEA assisted operations.

The IAEA safety standards also form the basis for the IAEA's safety review services, and they are used by the IAEA in support of competence building, including the development of educational curricula and training courses.

International conventions contain requirements similar to those in the IAEA safety standards and make them binding on contracting parties. The IAEA safety standards, supplemented by international conventions, industry standards and detailed national requirements, establish a consistent basis for protecting people and the environment. There will also be some special aspects of safety that need to be assessed at the national level. For example, many of the IAEA safety standards, in particular those addressing aspects of safety in planning or design, are intended to apply primarily to new facilities and activities. The requirements established in the IAEA safety standards might not be fully met at some existing facilities that were built to earlier standards. The way in which IAEA safety standards are to be applied to such facilities is a decision for individual States.

The scientific considerations underlying the IAEA safety standards provide an objective basis for decisions concerning safety; however, decision makers must also make informed judgements and must determine how best to balance the benefits of an action or an activity against the associated radiation risks and any other detrimental impacts to which it gives rise.

DEVELOPMENT PROCESS FOR THE IAEA SAFETY STANDARDS

The preparation and review of the safety standards involves the IAEA Secretariat and five Safety Standards Committees, for emergency preparedness and response (EPReSC) (as of 2016), nuclear safety (NUSSC), radiation safety (RASSC), the safety of radioactive waste (WASSC) and the safe transport of radioactive material (TRANSSC), and a Commission on Safety Standards (CSS) which oversees the IAEA safety standards programme (see Fig. 2).

All IAEA Member States may nominate experts for the Safety Standards Committees and may provide comments on draft standards. The membership of the Commission on Safety Standards is appointed by the Director General and includes senior governmental officials having responsibility for establishing national standards.

A management system has been established for the processes of planning, developing, reviewing, revising and establishing the IAEA safety standards.

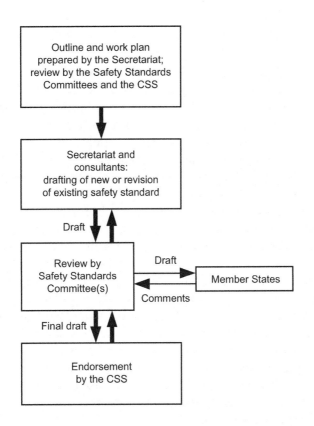

FIG. 2. The process for developing a new safety standard or revising an existing standard.

It articulates the mandate of the IAEA, the vision for the future application of the safety standards, policies and strategies, and corresponding functions and responsibilities.

INTERACTION WITH OTHER INTERNATIONAL ORGANIZATIONS

The findings of the United Nations Scientific Committee on the Effects of Atomic Radiation (UNSCEAR) and the recommendations of international expert bodies, notably the International Commission on Radiological Protection (ICRP), are taken into account in developing the IAEA safety standards. Some safety standards are developed in cooperation with other bodies in the United Nations system or other specialized agencies, including the Food and Agriculture Organization of the United Nations, the United Nations Environment Programme, the International Labour Organization, the OECD Nuclear Energy Agency, the Pan American Health Organization and the World Health Organization.

INTERPRETATION OF THE TEXT

Safety related terms are to be understood as defined in the IAEA Safety Glossary (see https://www.iaea.org/resources/safety-standards/safety-glossary). Otherwise, words are used with the spellings and meanings assigned to them in the latest edition of The Concise Oxford Dictionary. For Safety Guides, the English version of the text is the authoritative version.

The background and context of each standard in the IAEA Safety Standards Series and its objective, scope and structure are explained in Section 1, Introduction, of each publication.

Material for which there is no appropriate place in the body text (e.g. material that is subsidiary to or separate from the body text, is included in support of statements in the body text, or describes methods of calculation, procedures or limits and conditions) may be presented in appendices or annexes.

An appendix, if included, is considered to form an integral part of the safety standard. Material in an appendix has the same status as the body text, and the IAEA assumes authorship of it. Annexes and footnotes to the main text, if included, are used to provide practical examples or additional information or explanation. Annexes and footnotes are not integral parts of the main text. Annex material published by the IAEA is not necessarily issued under its authorship; material under other authorship may be presented in annexes to the safety standards. Extraneous material presented in annexes is excerpted and adapted as necessary to be generally useful.

CONTENTS

1. INTRODUCTION

BACKGROUND

1.1.   Requirements for the design of nuclear installations are established in IAEA Safety Standards Series Nos SSR-2/1 (Rev. 1), Safety of Nuclear Power Plants: Design [1]; SSR-3, Safety of Research Reactors [2]; and SSR-4, Safety of Nuclear Fuel Cycle Facilities [3].

1.2.   IAEA Safety Standards Series No. SSR-1, Site Evaluation for Nuclear Installations [4], establishes requirements for the external hazards that need to be considered in the evaluation of sites for nuclear installations.

1.3.   This Safety Guide provides specific recommendations on the design of nuclear installations to cope with the effects of external events[1], excluding earthquakes.

1.4.   This Safety Guide incorporates progress in the design of nuclear installations and in regulatory practice in States, taking into account lessons identified from extreme external events, feedback from safety review missions and the results of recent research on the effects of all external events excluding earthquakes. This Safety Guide provides new or updated recommendations on the following topics:

(a)   General concepts and application of safety criteria relating to the following:
   (i)   The design of structures, systems and components (SSCs) for protection against external events;
   (ii)   Load combinations;
   (iii)   The establishment of acceptance criteria.

---

[1]   An external event is an event that is unconnected with the operation of a facility or the conduct of an activity that could have an effect on the safety of the facility or activity [5]. Such events normally originate outside the site, and their effects on the nuclear installation need to be considered. Events originating on the site but outside safety related buildings are treated the same as off-site external events. External events could be of natural or human induced origin and are identified and selected for design purposes during the site evaluation process.

(b) Safety analysis for design basis external events[2] and beyond design basis external events[3].

(c) The design basis for each external event.

(d) Categorization of SSCs.

(e) Design and qualification methods and means of protection.

(f) Application of the management system.

1.5. Recommendations on site evaluation, focusing on the assessment of hazards, are provided in IAEA Safety Standards Series Nos SSG-9 (Rev. 1), Seismic Hazards in Site Evaluation for Nuclear Installations [6]; SSG-18, Meteorological and Hydrological Hazards in Site Evaluation for Nuclear Installations [7]; SSG-21, Volcanic Hazards in Site Evaluation for Nuclear Installations [8]; and NS-G-3.1, External Human Induced Events in Site Evaluation for Nuclear Power Plants [9].

1.6. Other Safety Guides provide recommendations on the design of nuclear installations for protection against the effects of hazards other than those addressed here and, as such, are complementary to this Safety Guide, particularly for the consideration of combinations of hazards and effects. These include IAEA Safety Standards Series Nos SSG-64, Protection against Internal Hazards in the Design of Nuclear Power Plants [10], and SSG-67, Seismic Design for Nuclear Installations [11].

1.7. This Safety Guide supersedes IAEA Safety Standards Series No. NS-G-1.5, External Events Excluding Earthquakes in the Design of Nuclear Power Plants[4].

---

[2] A design basis external event is an external event, or a combination of external events, that is considered in the design basis of all or any part of a facility [5]. Design basis external events are independent of the installation layout.

[3] The term 'beyond design basis external event' is used to indicate a level of external hazard exceeding those hazard levels considered for design, derived from the hazard evaluation for the site. The purpose of identifying beyond design basis external events is to ensure that the design incorporates features to enhance the capability of the installation to withstand such events. In addition, the identification of such events is used in evaluating the margins that exist in the design and in identifying potential cliff edge effects.

[4] INTERNATIONAL ATOMIC ENERGY AGENCY, External Events Excluding Earthquakes in the Design of Nuclear Power Plants, IAEA Safety Standards Series No. NS-G-1.5, IAEA, Vienna (2003).

OBJECTIVE

1.8. The objective of this Safety Guide is to provide recommendations on the design of nuclear installations for protection against the effects of external events excluding earthquakes. In particular, this Safety Guide is intended to provide recommendations on engineering related matters in order to meet the applicable safety requirements established in SSR-2/1 (Rev. 1) [1], SSR-3 [2] and SSR-4 [3] for the protection of nuclear installations against such events.

1.9. This Safety Guide also provides methods and procedures for defining an appropriate design for a nuclear installation on the basis of the site hazard evaluation and the layout of the installation. The aim is to provide guidance for the design of the installation, in particular for the protection of SSCs important to safety against design basis external events to ensure the safety of the installation.

1.10. This Safety Guide also provides recommendations on the selection of beyond design basis external events in order to check and verify margins and avoid cliff edge effects[5].

1.11. This Safety Guide is intended for use by organizations involved in the design of nuclear installations against external events; in analysis, verification and review; and in the provision of technical support. It is also intended for use by regulatory bodies for the establishment of regulatory guides.

SCOPE

1.12. This Safety Guide is applicable to all types of nuclear installation, as defined in the IAEA Safety Glossary [5]. This includes nuclear power plants; research reactors (including subcritical assemblies) and any adjoining radioisotope production facilities; spent fuel storage facilities; facilities for the enrichment of uranium; nuclear fuel fabrication facilities; conversion facilities; facilities for the reprocessing of spent fuel; facilities for the predisposal management of radioactive waste arising from nuclear fuel cycle facilities; and nuclear fuel cycle related research and development facilities.

---

[5] A cliff edge effect is an instance of severely abnormal conditions caused by an abrupt transition from one status of a facility to another following a small deviation in a parameter or a small variation in an input value [5].

1.13. This Safety Guide is applicable to the design of new nuclear installations and the safety evaluation of existing nuclear installations in relation to the following external events:

(a) Human induced events:
  (i) Accidental aircraft crashes;
  (ii) Explosions (i.e. deflagrations and detonations) with or without fire and with or without secondary missiles[6];
  (iii) Release of corrosive or hazardous gases or liquids (e.g. asphyxiant, toxic) from off-site or on-site storage or during transport;
  (iv) Release of radioactive material from off-site sources or from the site;
  (v) Fire generated off the site or from on-site sources;
  (vi) Collision of ships or floating debris with safety related structures, such as water intakes or structures associated with the ultimate heat sink;
  (vii) Collision of vehicles with SSCs;
  (viii) Electromagnetic interference from off-site or on-site sources;
  (ix) Floods resulting from the rupture of external pipes;
  (x) Any combination of the above resulting from a common initiating event, for example an explosion with fire and a release of hazardous gases and smoke.
(b) Natural events:
  (i) Floods due to events such as tides; tsunamis; seiches; storm surges; wind generated waves; precipitation causing flooding of nearby rivers and streams; dam forming and dam failures; bores and mechanically induced waves; channel migration; and high groundwater levels;
  (ii) Extreme meteorological conditions (of temperature, snow, hail, frost, subsurface freezing and drought);
  (iii) Extreme winds, including straight line winds, winds due to tropical storms (e.g. cyclones, hurricanes, typhoons) and tornadoes;
  (iv) Dust and sandstorms;
  (v) Lightning;
  (vi) Volcanism;
  (vii) Biological phenomena;

---

[6] The term 'missile' is used to describe a mass that has kinetic energy and has left its design location. This term is used to describe a moving object in general; military missiles, whether explosive or not, are specifically excluded from consideration. In general, military projectiles have velocities higher than Mach 1 and are therefore usually beyond the range of applicability of the techniques described in this Safety Guide. However, for non-explosive military projectiles with characteristics lying within the quoted ranges of applicability, the techniques described may be used.

(viii) Collision of floating objects (e.g. ice, logs) with safety related structures such as water intakes and components of the ultimate heat sink;

(ix) Geotechnical hazards (not associated with seismic loads);

(x) Any combination of the above.

This list might not be exhaustive for every site. Consequently, any other external events that are relevant for the site should be identified and evaluated. The hazards associated with external events may be affected by changes that have occurred since the siting process, as indicated in SSR-1 [4]. Such changes are considered in periodic safety reviews [12]. However, the hazard definition and protection concept also need to be reviewed following significant events that identify shortfalls in current knowledge and understanding, as well as when other significant new information has become available.

1.14. Throughout this publication, the term 'external events' always excludes earthquakes, for which recommendations are provided in SSG-67 [11].

1.15. The external human induced events considered in this Safety Guide are of accidental origin. Actions relating to sabotage are outside the scope of this Safety Guide, although some of the safety measures described might also be consistent with the needs of nuclear security. Specific guidance on the protection of nuclear power plants against sabotage is provided in Ref. [13].

1.16. The recommendations in this Safety Guide apply to all types of nuclear installation (see para. 1.12), including reactor types other than water cooled reactors at stationary nuclear power plants. The recommendations provided for nuclear power plants are also applicable to other nuclear installations through the application of a graded approach. Section 6 provides recommendations on the graded approach that should be followed for different types of nuclear installation.

1.17. This Safety Guide is mainly focused on the design stage of a new nuclear installation. However, the recommendations are also applicable in the re-evaluation of existing installations and in the periodic safety review described in IAEA Safety Standards Series No. SSG-25, Periodic Safety Review for Nuclear Power Plants [12].

STRUCTURE

1.18. The general concepts and application of safety criteria to the design of nuclear installations for protection against external events are presented in Section 2, including the relevant safety requirements; SSCs to be protected against external events; and recommendations on design and evaluation for design basis external events and beyond design basis external events and for the determination of adequate margins. Recommendations on the derivation of design parameters from the site evaluation, and on the overall design approach and evaluation of beyond design basis external events, are provided in Section 3. Recommendations on installation layout and the approach to the design of buildings are provided in Section 4. Recommendations on specific external events are provided in Section 5. Section 6 provides recommendations on design provisions for nuclear installations other than nuclear power plants using a graded approach. Section 7 provides recommendations on the application of the management system to the design of nuclear installations for protection against external events.

2. GENERAL CONCEPTS AND APPLICATION OF SAFETY CRITERIA TO THE DESIGN OF NUCLEAR INSTALLATIONS AGAINST EXTERNAL EVENTS

REQUIREMENTS FOR SITE EVALUATION

2.1. In accordance with Requirement 7 of SSR-1 [4], proposed sites for a nuclear installation are required to be evaluated in terms of natural and human induced external events, with emphasis on the frequency of exceedance and severity of the events. Hazards should be evaluated using deterministic and, as far as practicable, probabilistic methods, taking into account best practice. Potential combinations of such events are also required to be considered in the design of the installation (see para. 4.20 of SSR-1 [4]).

DESIGN REQUIREMENTS FOR NUCLEAR INSTALLATIONS

2.2. The requirements relevant to the design of nuclear power plants are established in SSR-2/1 (Rev. 1) [1]. For the design of research reactors and nuclear fuel cycle facilities, relevant requirements are established in SSR-3 [2] and SSR-4 [3], respectively. All these Safety Requirements publications stress the importance of applying a graded approach. Where no specific safety requirements for design have been established for a particular type of nuclear installation, the requirements established in SSR-2/1 (Rev. 1) [1], SSR-3 [2] and SSR-4 [3] should be applied, as far as practicable, using the graded approach described in Section 6.

External hazards

2.3. With regard to considering external hazards in the design of nuclear power plants, SSR-2/1 (Rev. 1) [1] states (footnotes omitted in paras 5.17 and 5.21):

"**Requirement 17: Internal and external hazards**

"**All foreseeable internal hazards and external hazards, including the potential for human induced events directly or indirectly to affect the safety of the nuclear power plant, shall be identified and their effects shall be evaluated. Hazards shall be considered in designing the layout of the plant and in determining the postulated initiating events and generated loadings for use in the design of relevant items important to safety for the plant.**

"5.15A. Items important to safety shall be designed and located, with due consideration of other implications for safety, to withstand the effects of hazards or to be protected, in accordance with their importance to safety, against hazards and against common cause failure mechanisms generated by hazards.

"5.15B. For multiple unit plant sites, the design shall take due account of the potential for specific hazards to give rise to impacts on several or even all units on the site simultaneously.

. . . .

"5.17. The design shall include due consideration of those natural and human induced external events (i.e. events of origin external to the plant) that have been identified in the site evaluation process. Causation and likelihood shall be considered in postulating potential hazards. In the short term, the safety of the plant shall not be permitted to be dependent on the availability of off-site services such as electricity supply and firefighting services. The design shall take due account of site specific conditions to determine the maximum delay time by which off-site services need to be available.

....

"5.19. Features shall be provided to minimize any interactions between buildings containing items important to safety (including power cabling and control cabling) and any other plant structure as a result of external events considered in the design.

....

"5.21. The design of the plant shall provide for an adequate margin to protect items important to safety against levels of external hazards to be considered for design, derived from the hazard evaluation for the site, and to avoid cliff edge effects.

"5.21A. The design of the plant shall also provide for an adequate margin to protect items ultimately necessary to prevent an early radioactive release or a large radioactive release in the event of levels of natural hazards exceeding those considered for design, derived from the hazard evaluation for the site."

Similar provisions for considering external hazards are established in Requirement 19 of SSR-3 [2] for the design of research reactors and in Requirement 16 of SSR-4 [3] for the design of nuclear fuel cycle facilities.

Engineering design rules

2.4. SSR-2/1 (Rev. 1) [1] states:

"Requirement 18: Engineering design rules

"The engineering design rules for items important to safety at a nuclear power plant shall be specified and shall comply with the relevant national or international codes and standards and with proven engineering practices, with due account taken of their relevance to nuclear power technology.

"5.23. Methods to ensure a robust design shall be applied, and proven engineering practices shall be adhered to in the design of a nuclear power plant to ensure that the fundamental safety functions are achieved for all operational states and for all accident conditions."

Similar provisions for engineering design rules and proven engineering practices are established in Requirement 13 of SSR-3 [2] for the design of research reactors and in Requirement 12 of SSR-4 [3] for the design of nuclear fuel cycle facilities.

Design extension conditions

2.5. SSR-2/1 (Rev. 1) [1] states:

"**Requirement 20: Design extension conditions**

"**A set of design extension conditions shall be derived on the basis of engineering judgement, deterministic assessments and probabilistic assessments for the purpose of further improving the safety of the nuclear power plant by enhancing the plant's capabilities to withstand, without unacceptable radiological consequences, accidents that are either more severe than design basis accidents or that involve additional failures. These design extension conditions shall be used to identify the additional accident scenarios to be addressed in the design and to plan practicable provisions for the prevention of such accidents or mitigation of their consequences.**"

The same provisions for design extension conditions are established in Requirement 22 of SSR-3 [2] for the design of research reactors and in Requirement 21 of SSR-4 [3] for the design of nuclear fuel cycle facilities.

Heat transfer to an ultimate heat sink

2.6. SSR-2/1 (Rev. 1) [1] states that with respect to nuclear power plants:

"**Requirement 53: Heat transfer to an ultimate heat sink**

"**The capability to transfer heat to an ultimate heat sink shall be ensured for all plant states.**

. . . .

"6.19B. The heat transfer function shall be fulfilled for levels of natural hazards more severe than those considered for design, derived from the hazard evaluation for the site."

There are no equivalent requirements in SSR-3 [2] or SSR-4 [3] in relation to the design of research reactors or nuclear fuel cycle facilities. Consequently, where the design of other nuclear installations needs to include the capability to transfer heat to an ultimate heat sink, a graded approach should be applied using the requirements established in SSR-2/1 (Rev. 1) [1] as a starting point.

Control room

2.7. SSR-2/1 (Rev. 1) [1] states:

> **"Requirement 65: Control room**
>
> **"A control room shall be provided at the nuclear power plant from which the plant can be safely operated in all operational states, either automatically or manually, and from which measures can be taken to maintain the plant in a safe state or to bring it back into a safe state after anticipated operational occurrences and accident conditions.**
>
> ....
>
> "6.40A. The design of the control room shall provide an adequate margin against levels of natural hazards more severe than those considered for design, derived from the hazard evaluation for the site."

Similar provisions for the control room are established in Requirement 53 of SSR-3 [2] for the design of research reactors; however, there are no equivalent requirements in SSR-4 [3] for the design of nuclear fuel cycle facilities.

OTHER ASPECTS OF DESIGN AGAINST EXTERNAL EVENTS

2.8. The hazard evaluation (see para. 2.1) should provide the following information:

(a) The severity levels of hazards and the annual frequencies of exceedance;
(b) Descriptions of the hazard evaluation methods and the elements and parameters of importance (including screening methods, results and uncertainties);
(c) The assumptions made in the hazard evaluation process.

This information should be communicated to the organization responsible for the design of the nuclear installation.

2.9. Two levels of external hazard should be considered in the design and evaluation of the SSCs important to safety in a nuclear installation that are subjected to external events. The first level is the design basis external event, and the second level is the beyond design basis external event. The frequencies of exceedance of design basis external events should be low enough that the design measures applied will ensure a high degree of protection with respect to external hazards.[7] It should be specified whether the frequencies of exceedance of the design basis external events refer to the mean, the median or a specific percentile.

2.10. The organization responsible for the design of the nuclear installation should define the loading conditions for design basis external events to be used in the design of SSCs and the loading conditions to be used in the evaluation of SSCs for beyond design basis external events. These loading conditions should be determined using the information from the hazard evaluation.

2.11. A design with an adequate margin, as required by para. 5.21 of SSR-2/1 (Rev. 1) [1], is achieved through conservative design approaches, duly taking into account the variability and uncertainties of the different methods, data, assumptions and rules used. The aim is to ensure that SSCs have the capability to perform safely even in situations more severe than those postulated in the design basis, without the occurrence of cliff edge effects. A source of margin is provided in the design of SSCs for a wide range of internal and external extreme loads. The governing loads for some SSCs could be different from, for example, the pressure and other environmental loads due to accident conditions, aircraft crashes, tornadoes, pipe breaks or seismic loads.

2.12. With regard to the design of SSCs, adequate margins are derived from the method used to define the loading conditions and from compliance with stress limits defined by the design codes and manufacturing codes. Nuclear design codes and standards implicitly or explicitly yield the margin achieved in the design process for individual SSCs. The margin for individual SSCs (i.e. the margin that results from the consideration of a variety of load cases) or for the nuclear installation as a whole should be achieved through the chain of steps from specifying the loading parameters to defining and achieving the performance acceptance criteria for the SSCs.

---

[7] In many States, a target frequency of exceedance of 10^{-4} per year or less is used for design basis external events for natural hazards.

2.13. For the purpose of this Safety Guide, the term 'adequate margin' refers to the following:

(a) The overall capacity of the nuclear installation to withstand the loading conditions of design basis external events;
(b) The capacity of individual SSCs to perform their intended function when subjected to the loading conditions of design basis external events;
(c) The avoidance of any cliff edge effects due to beyond design basis external events.

2.14. A design basis external event and its corresponding loading conditions should be defined conservatively in terms of the associated margins, to take into account any uncertainties.

2.15. Beyond design basis external events can be defined in the following ways:

(a) By adopting a lower annual frequency of exceedance than that specified for the design basis external event.
(b) By adopting a higher amplitude in the design basis external event loading conditions for all SSCs important to safety, or for a subset of SSCs ultimately necessary to prevent an early radioactive release or a large radioactive release. One approach is to add a factor of conservatism to the design basis external event loading conditions for such SSCs.

2.16. When considering a beyond design basis external event and following a best estimate approach, values of external event parameters causing cliff edge effects should be established. Adequate margins to these values should be demonstrated.

2.17. An assessment of adequate margins should be performed to determine either of the following:

(a) The level of the loading conditions at which the applicable safety functions for the installation or the function of an SSC important to safety would be compromised. This assessment process should include the identification of weak links and areas of improvement for engineering design. The margin assessment should also identify the potential for cliff edge effects due to external events and estimate their probability of occurrence.
(b) The level of the loading conditions at which there is high confidence that the applicable safety functions for the installation would be fulfilled and at which there would be no cliff edge effects as a result of slightly greater loading conditions.

These two approaches represent the probabilistic approach and the deterministic approach, respectively. The probabilistic approach should provide the best estimate of the level of loading conditions at which the applicable safety functions for the installation would continue to be fulfilled. The deterministic approach should provide conservative values at which there is high confidence that the applicable safety functions for the installation would be fulfilled.

2.18. In the probabilistic approach, the best estimate value should be defined by the mean or median values of the loading conditions. The best estimate value should be calculated using full probabilistic models of the loading conditions, the response of the installation and the capacity of SSCs important to safety. The best estimate value should be convolved over the range of values or derived as a point estimate using a simple best estimate model in which the loading condition is defined as the mean or median value and all installation parameters are assigned their best estimate values.

2.19. In the deterministic approach, a metric should be defined for the margin assessment. One such approach uses the 'high confidence of low probability of failure' capacity. This approach is commonly used in seismic margin assessments, and further recommendations are provided in SSG-67 [11] and IAEA Safety Standards Series No. NS-G-2.13, Evaluation of Seismic Safety for Existing Nuclear Installations [14].

2.20. The margins to be defined for various external hazards depend on the attributes of these hazards. Some attributes potentially increase the severity or the consequences of external events, while others potentially mitigate the effects of external events. The following factors should be considered in defining adequate margins:

(a) Factors that potentially make the effects of external events on a nuclear installation (especially a nuclear power plant) more severe and more uncertain:
  (i) The potential for causing cliff edge effects.
  (ii) Uncertainties in the hazard evaluation (e.g. database issues, such as completeness and constraints for maximum values).
  (iii) Limited experience of specific external events or in relation to specific nuclear installations (i.e. maturity of subject matter).
  (iv) The potential for combination with other external events and for interdependencies (e.g. high winds and flood; earthquake ground motion, fault displacement and tsunami).

(v)   The potential for an external event to cause an internal event (e.g. an earthquake causing an internal fire or flood).

(vi) The extent of common cause failure, such as the following:
— Simultaneous effects on all SSCs in one installation, on multiple units in a nuclear power plant or on multiple nuclear power plant sites;
— Potential compromise of redundancy of systems or of defence in depth;
— A simultaneous challenge to on-site and off-site emergency response measures.

(b) Factors that potentially mitigate the effects of external events on a nuclear installation:
(i)   The potential for advanced warning, such as the following:
— Warning time in hours for extreme weather conditions or external flooding (e.g. due to hurricane or cyclone, river flood, or tsunami from a distant source) and for dust in air intakes (e.g. due to volcano eruption or sandstorm);
— Warning time in minutes or less for extreme wind (e.g. tornado).

(ii) Having sufficient time to shut down the reactor (orderly or scram), noting that the shutdown state will need to be evaluated.

(iii) The extent of common cause failure, for example a limited spatial effect (footprint) due to tornado or aircraft crash.

2.21. In the evaluation of the safety of the nuclear installation in relation to beyond design basis external events, acceptance criteria applicable to the treatment of design extension conditions should be applied.

STRUCTURES, SYSTEMS AND COMPONENTS TO BE PROTECTED AGAINST EXTERNAL EVENTS

2.22. In the design and evaluation process for each individual external event, all SSCs important to safety that are affected by or exposed to the external event under consideration should be identified, including those SSCs whose failure could jeopardize SSCs important to safety. The list of the identified SSCs should include all equipment and any barriers or protective structures built to specifically address the external event.

2.23. The categorization of SSCs in relation to external events should follow the recommendations on seismic categorization provided in paras 3.31–3.40 of SSG-67 [11]. SSCs that are comparable to SSCs in seismic category 1 should be categorized as external event category 1. SSCs in external event category 1 should be designed to withstand the respective design basis external event, and an adequate margin should be provided to avoid cliff edge effects.

2.24. External event category 2 should be established for SSCs whose failure could jeopardize SSCs in external event category 1. Similar to seismic category 2, it should be demonstrated that SSCs in external event category 2 that have the potential to interact with SSCs in external event category 1 are effectively prevented from impairing those external event category 1 SSCs. SSCs in external event category 2 should be designed for the design basis external event, or it should be demonstrated that their failure will not impact the safety function of the external event category 1 SSCs.

2.25. External event category 3 should include all items that are not in external event categories 1 or 2. The items in external event category 3 should, at a minimum, be designed in accordance with the national approach to the external event design of high risk conventional (i.e. non-nuclear) installations.

DESIGN AND EVALUATION FOR DESIGN BASIS EXTERNAL EVENTS AND BEYOND DESIGN BASIS EXTERNAL EVENTS

2.26. The design of a nuclear installation for an external event should take into account all credible consequential effects of that event. External events can challenge the safety of a nuclear installation by different means, such as the following:

(a) Deterioration of site protection features (e.g. failure of human-made earthen structures, shielding walls or dykes);
(b) Deterioration of structural capabilities (e.g. leaktightness; structural integrity; support to equipment, components or distribution systems);
(c) Impairment of equipment operation;
(d) Impairment of redundancy of function due to common cause failure;
(e) Impairment of the capability of operating personnel;
(f) Unavailability of a heat sink;
(g) Unavailability of off-site power sources or off-site services and resources.

2.27. Having selected the external events to be considered for a particular site in accordance with the requirements established in SSR-1 [4], the effects of these events on the installation should be evaluated, including all credible secondary effects. The following should also be taken into account:

(a) When evaluating the effects of external events on the installation, it should be ensured that realistic and credible scenarios are identified and covered by a conservative scenario.
(b) For evaluations of beyond design basis external events, deterministic and — as far as practicable — probabilistic methods should be used to assess safety margins.

2.28. With regard to the design of nuclear power plants, SSR-2/1 (Rev. 1) [1] states:

"Requirement 24: Common cause failures

"The design of equipment shall take due account of the potential for common cause failures of items important to safety, to determine how the concepts of diversity, redundancy, physical separation and functional independence have to be applied to achieve the necessary reliability."

Similar provisions are established in Requirement 26 of SSR-3 [2] for the design of research reactors and in Requirement 23 of SSR-4 [3] for the design of nuclear fuel cycle facilities.

2.29. With regard to the design of nuclear power plants, Requirement 25 of SSR-2/1 (Rev. 1) [1] states that **"The single failure criterion shall be applied to each safety group incorporated in the plant design."** [8] Similar provisions are established in Requirement 25 of SSR-3 [2] for the design of research reactors and in para. 6.92 of SSR-4 [3] for the design of nuclear fuel cycle facilities.

2.30. For design, the single failure criterion is only capable of dealing with random failures. Therefore, the redundancy, which is the ultimate outcome of single failure criteria analysis, might be defeated by common cause failures

---

[8] In some States, the probability of occurrence of certain human induced events, such as external explosions or aircraft crashes, is considered very low, and passive components are usually assumed to be designed, manufactured, inspected and maintained to an extremely high quality. Therefore, the single failure non-compliance clause in para. 5.40 of SSR-2/1 (Rev. 1) [1] can be applied to the passive components. In some States, system outage due to repair, testing or maintenance, with its associated change in installation configuration, is considered one possible single failure mode in this context. Other States include the single failure criterion for all design basis external events.

associated with external events that have adverse effects over relatively large areas of the site.

2.31. Unless credible, a design basis external event or a beyond design basis external event should be considered in combination with other events that might occur independently, such as other external human induced events, natural phenomena, equipment failures and operator errors. Deterministic and probabilistic evaluations should be used for the determination and evaluation of suitable design combinations.

2.32. A loss of off-site power should be assumed to be coincident with a design basis external event or beyond design basis external event, unless a direct or indirect causal relationship can be excluded. In particular, for external events that are expected to affect the entire site and therefore give rise to a potential for a common cause failure, or for external events that might cause a turbine or reactor trip, a loss of off-site power should be combined with the design basis external event and the beyond design basis external event.

2.33. In the evaluation of design basis external events and beyond design basis external events that produce direct and indirect effects, the time delay between such effects should be taken into consideration in specifying how the direct and indirect effects are to be combined.

2.34. For phenomena of design basis external events and beyond design basis external events that are expected to develop slowly, the possibility of advance warnings and the implementation of precautions should be considered. In such cases, written procedures should be prepared to clearly define the actions to be taken once a warning is received. Consideration should be given to the immediate, medium term and long term effects of design basis external events and beyond design basis external events on off-site and on-site infrastructure and facilities, because non-nuclear on-site infrastructure and facilities may be damaged or destroyed by the external event (e.g. on-site roads or sea harbour landings for supply delivery).

2.35. Off-site infrastructure and assets that, under normal circumstances, are expected to provide various types of support to the nuclear installation might be unavailable. If these conditions could exist for a long period of time, the feasibility of providing support from off-site resources should be evaluated. Therefore, realistic assessments should be made of the ability to receive off-site support under extreme conditions in the site region. An adequate capacity of off-site infrastructure and assets should be ensured under such conditions, otherwise the provision of off-site support should be excluded from the safety analysis.

2.36. In general, any crediting of mitigatory actions that involve the support of off-site facilities should be based on the analysis of the specific beyond design basis external event and the particular site conditions and should include adequate margins to take account of uncertainties. When presuming the occurrence of external natural and human induced events, no credit for the support of off-site facilities, resources or services (e.g. equipment, electricity supply, firefighting services) is permitted in the short term (see para. 5.17 of SSR-2/1 (Rev. 1) [1]). Site specific conditions should also be taken into consideration when determining the time that would be needed for off-site facilities, resources and services to become available.

2.37. For an ultimate heat sink, the need for make-up of heat transport fluids and the possibility of using auxiliary junctions and injection points for heat removal systems should be examined. Where a limited quantity of heat transport fluids is stored on the site, the capability for make-up should be ensured by one of the following means:

(a) Protecting the make-up system from external events;
(b) Providing an adequate quantity of such fluids to allow time to repair the damaged part of the make-up system;
(c) Providing junctions and injection points to the system, adequately protected from external events, through which additional heat transport fluid can be injected from other on-site sources while the repair takes place.

2.38. Credit for the actions of operating personnel during or after a design basis external event, and the training necessary to perform these actions, should be determined on the basis of the specific external event and its anticipated effects on the site and SSCs. Impediments to the actions of operating personnel include the following:

(a) Lack of on-site communication;
(b) Lack of mobility due to soil failures on the site;
(c) Lack of specialized technical support needed to safely perform a recovery function;
(d) Inability to perform actions due to failures or malfunctions of SSCs;
(e) Inaccessibility of areas due to structural damage or changed environmental conditions.

2.39. No credit should be given for the actions of operating personnel to correct equipment failures, repair damage or suppress induced events (e.g. bushfire) as a consequence of a design basis external event or a beyond design basis external event, unless such actions can be safely and reliably accomplished within a time frame consistent with the complexity and difficulty of the actions. A considerable margin should be applied to take into account uncertainties; the time needed to diagnose the extent of failure and to develop or modify corrective procedures; and the possible unavailability of appropriate personnel or replacement parts.

DESIGN SAFETY FEATURES FOR DESIGN BASIS EXTERNAL EVENTS AND BEYOND DESIGN BASIS EXTERNAL EVENTS

2.40. In the design of a nuclear installation for protection against design basis external events, adequate robustness should be implemented to provide the installation with adequate margins for beyond design basis external events. In general, this capacity should be provided by a combination of high quality design; low sensitivity to variation in design parameters; and high and demonstrable conservatism in material selection, construction standards and quality assurance. An evaluation of the design conservatism should be carried out either with probabilistic tools or by a deterministic bounding analysis.

2.41. In designing a nuclear installation to withstand design basis external events, the systems of the installation should adhere to the single failure criterion for active components, which may be achieved by means of the redundancy of safety systems or trains in a system, taking due account of potential common cause failures. This criterion is also relevant for passive components, unless it has been demonstrated in the single failure analysis with a high level of confidence that a failure of a given passive component is very unlikely and that its function would remain unaffected by the design basis external event (see para. 5.40 of SSR-2/1 (Rev. 1) [1] and para. 6.77 of SSR-3 [2]). The acceptance criteria used in relation to design basis external events should be based on the acceptance criteria applicable for design basis accidents.

2.42. The protection of a nuclear installation against external events should be provided by one or more of the following approaches:

(a) The effects of an external event are reduced by means of a passive barrier (e.g. a 'dry site' (see paras 4.11–4.13), dykes or sea walls for floods; external shields for an aircraft crash; barriers for explosions).

(b)   Safety systems are designed to effectively resist the effects of external events by application of the concepts of diversity, redundancy, physical separation and functional independence (see Requirements 21 and 24 of SSR-2/1 (Rev. 1) [1], Requirements 26 and 27 of SSR-3 [2] and Requirement 23 of SSR-4 [3]).

(c)   SSCs are designed to withstand the external event loading conditions.

(d)   Administrative measures are implemented, such as the establishment and enforcement of no-fly zones.

The justification of the approaches used to protect the nuclear installation should identify the rationale for the choice of methods and include a demonstration of the reliability of these methods. Administrative measures as a replacement for passive or active protection should be avoided, as far as reasonably practicable.

2.43.   Special provisions against common cause failure should be made for large and extensive systems, such as pump houses, cooling towers, systems used to transport heat to the ultimate heat sink, or long piping systems with large ring main systems. A combination of the following protective measures should be implemented:

(a)   Adequate redundancy of items important to safety. The level of redundancy should be an outcome of the application of the single failure criterion to the design.

(b)   Adequate spatial separation between redundant components. This should aim to prevent common cause failures from localized external events (e.g. missile impact) and from interactions due to the failure of one system being a source of failure of another. A detailed analysis of the areas of influence or expected damage from the design basis external event and the beyond design basis external event should be carried out for the purpose of applying the concept of physical separation.

(c)   Diversity in the redundant components. For external event scenarios with a potential for common cause failures, the benefits of diversity should be evaluated with care. Diversity should be combined with separation when possible.

2.44.   The design of a new nuclear installation should represent the best balance of system layout, safety aspects (system and nuclear installation) and operational aspects, taking into account external events.

2.45.   For design modifications to an existing nuclear installation to specifically address changes in the assessment of the site specific hazard, design options such as relocating redundant systems or elements of systems might be limited. In such cases, additional safety measures should be implemented, if reasonably practicable, in the form of barriers or of retrofitting portions of systems to achieve the necessary functional capacity. Options that should be considered include installing additional permanent equipment and making available (on the site and/or off the site) non-permanent (i.e. temporary) equipment, which may be mobilized if needed. The additional permanent and non-permanent equipment should be categorized in accordance with paras 2.23–2.25 to ensure that it functions when needed.

2.46. The following should be considered in the design of a nuclear installation against design basis external events:

(a) The main control room, the supplementary control room and other locations (e.g. compartments, rooms, facilities) necessary for meeting operating requirements should be accessible during and after a design basis external event.
(b) Items associated with defence in depth level 3 and defence in depth level 4 should not be impaired as a consequence of a design basis external event.
(c) Systems not protected against design basis external events should be assumed to be either operable or non-operable, depending on which status provides the more conservative scenario in the design of protective measures against the design basis external event.
(d) The on-site mobility of personnel and equipment, if needed after the occurrence of a design basis external event, should be verified.

2.47. The following should be considered in the design of a nuclear installation against beyond design basis external events:

(a) The main control room, the supplementary control room, and any locations (e.g. compartments, rooms, facilities) necessary for responding to a beyond design basis external event, should be accessible during and after the event.
(b) Systems not protected against beyond design basis external events should be assumed to be operable or non-operable, depending on which status provides the more conservative scenario in the evaluation of protective measures against the beyond design basis external event. Provided there is adequate justification, the non-operability of unprotected systems may be assumed.
(c) Systems not protected against beyond design basis external events and that are not important to safety should be designed such that their failure due to such an event will not jeopardize SSCs important to safety.

(d) The on-site mobility of personnel and equipment, if needed after the occurrence of a beyond design basis external event, should be verified.

2.48. Provisions in the design to protect the installation against design basis external events and beyond design basis external events should not impair the ability of the installation to withstand other design basis events or otherwise affect safety related operating procedures.

2.49. If any SSC (including the complete nuclear installation) incorporates seismic isolation in its design, it should be demonstrated that the response of the SSC to other external hazards is not adversely affected by this design approach.

ADMINISTRATIVE MEASURES

2.50. Administrative measures for design basis external events and beyond design basis external events include procedures and protocols that help ensure that the safety objectives for the nuclear installation are met. Administrative measures, in conjunction with other measures, should be developed as part of the protection scheme for each external event, as appropriate. Administrative measures taken prior to an external event should be based on the considerations presented in para. 2.34. When applicable, this should include measures such as warnings and preparations for tsunamis, hurricanes and tornadoes and for the release of hazardous gases and liquids. Furthermore, procedures and protocols should be established to prevent hazardous situations, for example a no-fly zone within a given radius around the nuclear installation site, restrictions on the storage of on-site materials that could become missiles and restrictions on the storage of combustible materials on the site.

2.51. The effectiveness of administrative measures is strongly dependent on their enforcement, particularly when different administrations are involved (i.e. administrations outside of the operating organization of the nuclear installation). Administrative measures should be used in conjunction with other measures: to the extent possible, administrative measures should act as an additional layer of defence. The reliability and effectiveness of such measures should be carefully evaluated on a periodic basis.

3. DESIGN BASIS FOR EXTERNAL EVENTS

DERIVATION OF THE DESIGN BASIS FROM THE SITE HAZARD EVALUATION

3.1. The end products of the site hazard evaluation are specified in para. 2.8. Adequate communication should be maintained between the individuals conducting the site hazard evaluation and those performing the design of the nuclear installation to ensure that adequate information and data are available to develop the loading conditions for external events. The information and data should be transparent and understandable so that the development of the loading conditions is also transparent and understandable to all stakeholders.

3.2. The design process should include the provision of information to the site hazard evaluation regarding the derivation of design basis external events and beyond design basis external events, including the appropriate level of annual frequency of exceedance to be considered.

3.3. Screening is a part of the hazard analysis in site evaluation (see Requirement 6 of SSR-1 [4]). For human induced external events, screening by physical distance (e.g. using a screening distance value) as well as by severity or probability of occurrence[9] should be undertaken. When a screening probability level approach is used for screening purposes, the level of annual frequency of exceedance to be considered should be agreed before the site hazard evaluation.

3.4. A feedback process between the site hazard evaluation and the installation design should be implemented. This process should include feedback on hazard parameters and loading conditions and on the results of the site hazard screening process.

3.5. The general approach in the design is to establish the design loading conditions through a combination of deterministic and probabilistic methods and

---

[9] In some States, a value for the probability of 10^{-7} per reactor-year is used in the design of new facilities as an acceptable limit on the probability value for interacting events that have serious radiological consequences. This is considered a conservative value for the screening probability level if applied to all events of the same type (e.g. all aircraft crashes, all explosions). Some initial events may have very low limits on their acceptable probability and need to be considered in isolation [9].

to proceed with the design in a deterministic manner. A detailed discussion of the appropriate approaches is contained in paras 2.19–2.27 of SSG-18 [7].

3.6. Even if the combined deterministic and probabilistic approaches identify a specific loading condition as a potential design basis external event, it may still be excluded from further analysis if it is shown that the corresponding loading conditions are completely bounded by the loading conditions of other design basis events that have already been considered (see para. 4.18 of SSR-1 [4]). However, the screened-out hazard should still be kept in the design basis to ensure that potential engineering and administrative measures taken for the bounding case are indeed valid.

3.7. When the hazard is defined in a probabilistic context, the site hazard should be analysed and presented in a set of hazard curves. At the design stage, the hazard curves or a single hazard value at a given annual frequency of exceedance should be used.

3.8. The final safety objective of the design basis selection is to keep the radiological risk due to external events as low as reasonably achievable and below any acceptable limits established by the regulatory body. For nuclear power plants, the mean annual core damage frequency, mean annual large early release frequency and/or the mean annual large release frequency need to be below the limits established by the regulatory body. To satisfy this objective, the following should be considered in the specification of design basis external events and beyond design basis external events:

(a) The likelihood of occurrence of such external events;
(b) The effects of such external events on SSCs important to safety;
(c) The consequences of the loading conditions on the ability of SSCs to meet performance requirements;
(d) The overall consequences on the installation with respect to the risk metrics.

3.9. An appropriate deterministic or probabilistic analysis should be performed at the level of detail necessary to demonstrate that the safety objective of the design has been met. For nuclear installations other than nuclear power plants, the graded approach should be applied as recommended in Section 6.

3.10. For each external event of interest, the possibility of the loading conditions creating a cliff edge effect should be assessed. The assessment should include the identification of the cliff edge effect (e.g. overtopping of a flood protection structure), the probability of the occurrence of such an effect, the consequences of the cliff edge effect on SSCs and on the installation, and methods of mitigating these effects.

OVERALL DESIGN APPROACH

3.11. The occurrence of any possible normal operational state of the nuclear installation (e.g. full power, hot shutdown, cold shutdown, refuelling outage, maintenance, repair) at the same time as a design basis external event or a beyond design basis external event should be considered.

3.12. The initial conditions of the installation for the design basis external event and beyond design basis external event should include the effects of causal and concomitant events such as the following:

(a) A causal event might occur when a storm causes damage off the site and on the site. For example, off-site damage occurs to a river dam, which results in the release of water that flows towards the installation. The on-site damage affects SSCs that protect the installation against flooding. Thus, the state of the installation at the time of flooding needs to be taken into account.
(b) A concomitant loading condition occurs for a typhoon where wind forces, extreme rainfall and storm surge occur simultaneously.

3.13. The initial conditions of the installation for the design basis external event and beyond design basis external event should also take into account the effects of measures that might lead to a change of state of the installation prior to the external event. One example of such measures is an advance warning leading to shutdown of the installation.

3.14. Systematic inspections by expert engineers, implemented as formal walkdowns of the installation, should be performed for new installations before commissioning to provide final verification of the design for external events (including for internal interactions through internal fire, flood, mechanical impact and electromagnetic interference), to verify that there are no unanticipated situations and to verify specific design features. The walkdown team should consist of experts in external events and in the design of structures and components, together with systems analysts and operating personnel, including maintenance personnel. Formal installation walkdowns should also be performed for existing installations to evaluate their robustness against external events. In the walkdowns, 'housekeeping' aspects should also be addressed, for example loose equipment and furniture, fastening of equipment (e.g. gas bottles, ladders), and transient fire loads.

DERIVATION OF DESIGN BASIS EXTERNAL EVENT LOADING CONDITIONS: GENERAL CONSIDERATIONS

3.15. The derivation of the design basis parameters and the relevant loading scheme for the selected design basis external events should be consistent with the level of detail necessary for the assessment of the design limit[10]. (The methods, models, calculations and testing used are closely tied to the acceptance criteria.)

3.16. The performance criteria should target, as appropriate, the overall and local structural integrity of SSCs (e.g. leaktightness; lack of perforation[11]; lack of scabbing[12]; operability of equipment, components and distribution systems) and the level of compliance associated with the design procedures to be applied (e.g. static; dynamic; linear; non-linear; one, two or three dimensional analyses).

3.17. Many of the loads corresponding to the external human induced events described in NS-G-3.1 [9] are impact or blast loads with a rapid rise time and a short duration that are characterized by limited energy or a defined momentum transfer. The loads are often localized, causing a substantial local effect on individual targets but little overall effect on massive structures. In such cases, load–time functions should be derived by analytical simulation or by experimentation, preferably using rigid targets.

3.18. If simplified engineering approaches are used in the design process, it should be confirmed that these approaches are appropriate for the design and include adequate conservatism.

3.19. Refined studies supported by numerical analyses and/or physical testing should be carried out for specific layout configurations. These include studies of grouping effects among cooling towers, dynamic amplification of tall and slender

---

[10] The design limit is an interpretation of acceptance criteria in terms of design parameters (e.g. elasticity, maximum crack opening, no buckling, maximum ductility).

[11] Perforation is the state when an impacting missile has passed completely through the target.

[12] Scabbing is the ejection of irregular pieces of the face of the target opposite the impact face as a result of a missile impact.

stacks, or — in the case of aircraft crashes — the dynamic interaction effects on large and flexible slabs.

3.20. A sensitivity analysis should be conducted on the input data.

DERIVATION OF DESIGN BASIS EXTERNAL EVENT AND BEYOND DESIGN BASIS EXTERNAL EVENT LOADING CONDITIONS FOR SPECIFIC EXTERNAL EVENTS

3.21. For each external event, the design basis external event and the beyond design basis external event is determined, starting with screening by magnitude and distance, screening by probability of occurrence, the categorization of SSCs, the definition of the loading conditions (parameters) associated with the design basis external event and the beyond design basis external event, the design and evaluation of the SSCs when subjected to the loading conditions, and the likelihood and consequences of failure of SSCs. For each external event of interest, the possibility of the external event loading conditions creating a cliff edge effect should be assessed.

EVALUATION OF BEYOND DESIGN BASIS EXTERNAL EVENTS: CLIFF EDGE EFFECTS

3.22. The design basis should avoid cliff edge effects, with any uncertainties associated with the design basis external event values being taken into account. The following information should be obtained regarding potential cliff edge effects:

(a) The external event for which a cliff edge effect could occur;
(b) The severity of the event at which the cliff edge effect could occur;
(c) The loading condition corresponding to triggering the cliff edge effect;
(d) The probability of occurrence of this hazard level.

3.23. To assess the margins and evaluate cliff edge effects, one of the following methods for defining the beyond design basis external event loading conditions should be used:

(a) Defining the beyond design basis external event loading conditions by applying a factor to the design basis external event loading conditions. This is similar to the approach for beyond design basis earthquake loading conditions for new nuclear installation designs (see para. 3.29 of SSG-67 [11]).

(b) Defining the beyond design basis external event loading conditions on the basis of the probabilistic hazard evaluation.

(c) Defining the beyond design basis external event loading conditions as the maximum credible hazard severity.

3.24. The definition of the beyond design basis external event loading conditions is inherently connected with the performance and acceptance criteria for SSCs and the nuclear installation. Similar to the approach to design extension conditions (see footnote 13 in SSR-2/1 (Rev. 1) [1], footnote 24 in SSR-3 [2] and footnote 25 in SSR-4 [3]), methodologies to evaluate beyond design basis external events may be performed by means of a best estimate approach (which is relaxed compared with design methods and acceptance criteria relating to material properties).

3.25. Two methodologies should be considered to evaluate how beyond design basis external events affect the safety of a nuclear installation, as follows:

(a) A probabilistic safety analysis of external events other than earthquakes that quantifies core damage frequency, fuel damage frequency, large early release frequency and large release frequency;[13]

(b) A margin assessment method that determines the level of severity of an external event at or below which there is a very high confidence that the core damage frequency or fuel damage frequency arising from the external event is acceptably low.

3.26. For nuclear power plants, the following, if identified in accordance with para. 2.22, should be checked against beyond design basis external events to demonstrate an adequate margin and avoidance of cliff edge effects:

(a) Items that are ultimately necessary to prevent an early radioactive release or a large radioactive release;

(b) Items that ensure heat transfer to an ultimate heat sink;

(c) Items that ensure the functions of the control room and, if the main control room is not available following the beyond design basis external event, items that ensure the functions of the supplementary control room.

---

[13] In addition to seismic hazards, probabilistic safety analyses have been performed for external hazards such as floods and extreme winds.

4. INSTALLATION LAYOUT AND DESIGN APPROACH

INSTALLATION LAYOUT

Physical separation

4.1. Many external events produce only localized effects; that is, they have an area of influence that does not extend to the whole installation. In such cases, if the physical separation of redundant independent safety systems (as required by Requirement 21 of SSR-2/1 (Rev. 1) [1], Requirement 27 of SSR-3 [2] and Requirement 23 of SSR-4 [3]) is sufficient, this physical separation can be used to achieve safety. When physical separation is credited, it should be demonstrated that the installation layout ensures that, outside the areas affected by the external event, there will always be items redundant to those affected.

4.2. If the area affected by an external event is limited but is not confined to a specific location, the recommendations provided in para. 4.1 should be met, with the assumption that the event could take place anywhere on the site.

4.3. When identifying areas within the installation that might be affected by an external event, the possible effects on any particular function caused by the impairment of a system might not be obvious.[14] Safety systems and their support systems should be evaluated as a whole.

4.4. When there is reliance on non-permanent equipment for the fulfilment of a safety function, normally in beyond design basis external event scenarios, the practicability of moving such equipment from storage locations (off the site and on the site) to connection points on the site should be demonstrated, taking into account the effects of the external event.

Protective structures

4.5. For most external events, building structures as normally designed for nuclear installations provide a good level of protection for SSCs important to

---

[14] For example, the repair time for a power line damaged by an event may determine the minimum amount of stored fuel needed for the diesel generators, if the supply of diesel oil from sources nearby cannot be guaranteed. As another example, the failure of a ventilation system due to an aircraft crash might lead to a temperature rise inside a building, which in turn might cause the malfunctioning of electronic and pneumatic equipment far away from the crash area.

safety. Structures of buildings important to safety are normally constructed of reinforced or prestressed concrete, with relatively thick external walls and few openings that, in turn, are closed by robust metal doors. Hence, from the perspective of designing against external events, it is good practice to locate items important to safety inside buildings rather than outdoors. This approach should be followed to the extent practicable.

4.6. There are instances in which locating an item important to safety inside a building structure is not practicable or even possible. This is the case, for example, for large tanks, induced draft cooling towers or containers storing flammable or explosive substances. In such cases, if sufficient physical separation between redundant items cannot be demonstrated, a protective structure designed against the applicable external events should be included in the layout.

4.7. For some external events, the loads will govern the design of a structure intended to withstand the event. This is usually the case, for example, for large aircraft impacts. In such cases, when the principle of physical separation cannot be used, the structure should be designed to withstand the external event, under the applicable acceptance criteria.

4.8. The principle of physical separation is not normally used for the containment building structure, since there is normally no redundant building. In such cases, the following layout approaches should be considered:

(a) Locating the primary containment within either a secondary containment or an external structure capable of withstanding postulated external events;
(b) Structurally decoupling inner structures from the external containment to reduce the external event loads on these structures and any items important to safety installed on them;
(c) Having a containment building with a low vertical profile to reduce the likelihood of aircraft impact;
(d) Providing redundant, physically separated safety trains inside the containment that are capable of withstanding postulated external events.

4.9. As a result of the installation layout, some structures can protect other structures and equipment against some external events, even though they have not been intentionally designed to do so. For example, a building may protect other structures from the effects of an explosion along a transport route if the building is located between those structures and the route.

4.10. Infill masonry walls on steel or concrete framed structures are not structurally effective against explosions. Continuous reinforced concrete walls and diaphragms should be considered for this type of load.

Dry site concept

4.11. The dry site concept (see para. 7.5 of SSG-18 [7]) should be the preferred layout approach for protection against floods. In accordance with this approach, the ground level around buildings and other components important to safety should be located above the estimated maximum level of the design basis flood.

4.12. When the dry site concept cannot be applied, the layout should include permanent flood barriers or protections, with carefully selected design bases that appropriately consider flood event characteristics (i.e. flood levels, as well as their duration and associated effects) and their uncertainties.

4.13. Irrespective of the existence of permanent flood barriers, it is considered good practice to place flood sensitive equipment important to safety inside watertight compartments of buildings or at elevations above the level of the flood. This practice should be followed as far as practicable.

Special consideration for layout in the design of a nuclear installation against external events

4.14. Attention should be paid to the possible failure (due to external events) of items not important to safety that might affect the ability of the installation to fulfil safety functions.

4.15. The design of roofs should not permit the buildup of snow, rain or ice exceeding the roof design loads. The layout should include provisions that take into account accidental clogging of systems for the discharge of surface and drainage wastewater.

4.16. Light or slender structures (e.g. light roofs, metal stacks) are the most sensitive to wind loads, and such structures should be avoided, as far as practicable, on sites prone to high winds. Wind sensitive structures not important to safety can be the source of wind-borne missiles that can affect items important to safety. Where metal towers and stacks are necessary, they should be designed to have low susceptibility to vortex shedding wind loads.

4.17. Some external events can be considered extreme events, which are more frequent than rare events. This is the case, for example, for wind loads that do not include tornado or hurricane conditions.[15] In such cases, external event loads should be combined with normal operational loads and with loads from other extreme events, using combination factors based on national practice. A combination of probable maximum storm surge with 10 year wind wave effects is one example.

4.18. Another factor that should be considered in the installation layout is ignition of gas or vapour accumulated in confined external areas, such as courtyards or alleys. Detonations under these conditions could result in high local overpressures. To reduce the likelihood of such events, the design should, as far as practicable, provide a compact layout devoid of long alleys and inner courtyards or provide adequate openings to prevent the accumulation of an explosive concentration of gases.

APPROACH TO STRUCTURAL DESIGN

General

4.19. The design of a building against a design basis external event is generally based on a deterministic analysis. In general, there are three ways of ensuring the safety functions are met:

(a) To design the building or a protective structure to withstand the loads resulting from the design basis external event, and thereby maintain the functionality of the equipment housed by the building;
(b) To provide a redundant building (located outside the area of influence of the external event) housing components and systems that can satisfactorily fulfil the safety functions assigned to the building (e.g. a redundant emergency diesel building);
(c) To limit the consequences of damage to the building, so that the applicable safety functions are fulfilled.

Paragraphs 4.20–4.41 refer mainly to option (a) above.

---

[15] In some States, the design basis wind speed for extreme events is determined on the basis of a 100 year return period (1% annual frequency of exceedance) [7], whereas rare design events are typically chosen with a much longer return period.

Loading derivation

4.20. For each external event to be considered in the design, hazard parameters should be used to derive design basis external event and beyond design basis external event parameters for the design and evaluation process. Care should be taken to maintain consistency between the results of the hazard analyses and the parameters to be used for design.

4.21. The derivation of the design basis parameters and the relevant loading scheme for the selected design basis external events should be consistent with the level of detail necessary for the design limit assessment (e.g. leaktightness, perforation) and with the level of accuracy associated with the design procedures to be applied (e.g. linear, non-linear, three dimensional, dynamic).

4.22. Computational tools allow full three dimensional fluid dynamics analysis to derive suitable load functions or to assess the capacity of structures. However, in some cases, simplified engineering approaches have been developed that are based on interpretation of test data or data from numerical analysis. An assessment of the assumptions and applicability of each technique should be carried out to check their appropriateness for the case of interest and their compatibility with the level of accuracy needed in the design.

4.23. For specific layout configurations, refined studies supported by numerical analyses or physical testing might be necessary. Typical examples are the grouping effects among cooling towers under wind load, the dynamic amplification of tall and slender stacks or — in the case of aircraft crashes — the dynamic interaction effects on large and flexible slabs.

Load combinations and acceptance criteria

4.24. Because of their infrequent nature and very short duration, statistically independent loadings from any single design basis external event are usually combined only with normal operational loads using unity load factors for all loadings. Multiple design basis external event loadings such as for aircraft crashes and explosions usually do not have to be combined. However, all effects from a single design basis external event should be properly time phased and combined, with due attention paid to the physical meaning of the combinations. Thus, for aircraft crashes, the various effects of the impact (e.g. missiles, induced vibrations, fuel fires) should be combined. Furthermore, when a causal relationship or correlations for simultaneous occurrence exist between events, the effects should be properly combined, as necessary. Recommendations on the approach to combined effects in relation to meteorological events and floods are provided in SSG-18 [7].

4.25. Acceptance criteria (e.g. functionality, leaktightness, stability) should be assessed in accordance with the category of the item (external event category 1 or external event category 2). Such criteria should be interpreted in design terms, leading to appropriate design limits (e.g. allowed leak rate, maximum crack opening, elasticity, maximum displacement).

4.26. For loads from design basis external events, the design should provide for essentially elastic structural behaviour. Limited inelastic behaviour may be permitted, as long as the overall structural response basically remains within the linear domain and the structure fulfils its safety function.

4.27. Where local inelastic deformation is intended to absorb the energy input of the load, inelastic behaviour should be considered acceptable for individual ductile structural elements (e.g. beams, slabs), provided the stability of the structure as a whole or the ability of the structural element to perform its safety function is not jeopardized.

4.28. Global structural inelastic behaviour may be considered acceptable for protective substructures (e.g. restraints, missile barriers) whose sole function is to provide protection against external event loads, as long as the displacements remain acceptable.

Procedures for structural design

4.29. Design procedures should be selected in accordance with the characteristics of the structure, the loading functions and the acceptance criteria to meet the design limits.

4.30. In the case of numerical models used in sequence (e.g. global–local), attention should be paid to consistency between different models to ensure that the final results are representative of the structural response and behaviour.

4.31. The level of detail of the numerical models should be sufficient to adequately represent the structural behaviour and should be consistent with the specified design limits. The methods used for modelling and analysis (of, for example, structural joints, steel rebars in reinforced concrete, structural interfaces and liners) should be reviewed and verified using other approaches, as necessary.

4.32. For numerical models, the finite element mesh should be validated for any specific load case to be analysed. To minimize the uncertainties associated with numerical approximations when using meshed models, analyses should be carried out, and the convergence of results should be checked, which can necessitate optimization of the mesh size. The discretization should be appropriate for the frequency content of the load. For short duration loads (typical in explosions), dedicated models, different from the traditional dynamic models used for seismic analysis, may be necessary.

Material properties

4.33. Material properties should be in agreement with material specifications and consistent with the construction and quality assurance procedures for the safety category of the particular item. For design basis purposes, minimum certified values of strength should be used, taking into account the ageing properties of materials.

4.34. In the design for impulsive loads (e.g. explosion, impact), credit may be taken for the increase in strength due to strain rate effects. Appropriate material models that are strain rate dependent should be used for impact analysis.

Equipment qualification

4.35. The equipment necessary for fulfilling safety functions during and after the occurrence of a design basis external event should be functionally qualified for the induced conditions, including vibration.

4.36. Equipment qualification for impact loads or impulse loads may be very different from qualification for earthquake induced vibrations, and therefore specific procedures should be selected, in accordance with the performance requirements (e.g. stability, integrity, functionality). The qualification conditions should be compared with the demand, usually represented by vibration, impact or impulse forcing functions at the anchor point to the structural support. Adequate safety margins should be provided, in accordance with the safety category of the item.

4.37. When applicable, equipment qualification should take into account the necessary functionality under conditions of dust, smoke, humidity, extreme temperatures, corrosive atmospheres and/or radioactive environments, combined with mechanical stress.

4.38. For some external events, such as those involving corrosive chemicals or biological phenomena, potential degradation might occur over a considerable time period. In such cases, the design might not need to include highly durable protective measures, provided the items subject to degradation can be inspected. The scope, frequency and methods of the inspection programme should be commensurate with the degradation rates. Any protective measures implemented should be able to be reapplied or repeated; alternatively, the design should include measures to inhibit, stop or reverse the degradation.

Interaction effects

4.39. External events can cause direct damage to the installation; such effects are called 'primary effects'. These primary effects might cause indirect damage by means of interaction mechanisms that can propagate the damage ('secondary effects'). These secondary effects should be included in the analysis of the events, as they might cause damage that could be comparable with (or even exceed) the damage caused by the primary effects. Secondary effects are explicitly addressed in the categorization of items (see paras 2.24 and 2.25).

4.40. In the case of building structures designed against an external event, the design should address the following effects on nearby SSCs:

(a) Failure and collapse of nearby structures;
(b) Secondary missiles generated from nearby SSCs;
(c) Flooding from failure of liquid retaining structures, not necessarily close to the building;
(d) Chemical releases from failure of containers or deposits of material;
(e) Secondary fires or explosions, as a result of failures in tanks containing flammable or explosive material;
(f) Electromagnetic interference generated by electrical faults.

4.41. Special emphasis should be given to potential interaction effects between components of the ultimate heat sink (e.g. failure of cooling towers and flooding from the ultimate heat sink basin) and other safety related structures.

APPROACH TO STRUCTURAL ASSESSMENT FOR BEYOND DESIGN BASIS EXTERNAL EVENTS

General

4.42. The rules for design basis external events and the rules for beyond design basis external events are different. The purpose of the structural assessment should be to show that the beyond design basis external event will not compromise the intended safety functions. For this purpose, the assessment for a beyond design basis external event may take credit for all safety margins intentionally or unintentionally introduced by the design process. Nonetheless, the design criteria should remain consistent with the safety requirements and consider adequate margins.

Loading derivation

4.43. For some external hazards, it may be possible to identify scenarios that are extremely unlikely yet still credible, and these scenarios could be selected as the basis for the beyond design basis external event. In such cases, the annual frequency of exceedance of the beyond design basis external event should be at least one order of magnitude less than that of the design basis external event.

4.44. For some external hazards, the approach above might also lead to non-credible scenarios. In such cases, a 'hazard-agnostic'[16] approach should be taken in which the beyond design basis external event is selected on the basis of an adequate margin with respect to the design basis external event.

4.45. Hazard parameters should be used as the basis for a set of beyond design basis parameters used for the structural assessment. In this process, consistency with the site evaluation hazard analysis should be maintained.

---

[16] In this Safety Guide, the term 'hazard-agnostic' is used to indicate a situation where the protection against a hazard is provided without complete knowledge of the size and frequency of the hazard. Generally, a standardized envelope design for external hazards constitutes a hazard-agnostic approach.

Load combinations and acceptance criteria

4.46. Beyond design basis external events should be considered as very infrequent, and corresponding loads should be combined only with normal operational loads.

4.47. During beyond design basis external events, widespread unrecoverable structural deformation within structures might occur. However, structural acceptance criteria should be established to ensure that all relevant safety functions are fulfilled.

Procedures for structural assessment

4.48. Procedures for structural assessment should normally be oriented to obtain realistic (i.e. median or best estimate) structural behaviour.

Material properties

4.49. Material properties should be consistent with the loading conditions induced by external events and should be in agreement with the material specifications and the construction and quality assurance procedures associated with the safety category of the particular item. In structural assessment for beyond design basis external events, it is normally acceptable to use less conservative values than those used for design (e.g. reduced material safety coefficients, use of values based on the results of tests on the actual materials used).

5. DESIGN PROVISIONS AGAINST EXTERNAL EVENTS

EXTERNAL FLOODS, INCLUDING TSUNAMIS

5.1. Recommendations on assessing the potential risk of flooding of a site due to diverse initiating causes and scenarios (and relevant potential combinations) are provided in SSG-18 [7]. The phenomena that should be considered include the following:

(a) Storm surges.
(b) Wind generated waves.

(c) Tsunamis.
(d) Seiches.
(e) Flooding of rivers and streams.
(f) Extreme precipitation events: local intense precipitation.
(g) Floods due to the sudden release of impounded water from the following:
    (i) Dams (i.e. due to dam failure);
    (ii) Ice dams;
    (iii) On-site water storage (ultimate heat sink).
(h) Bores and mechanically induced waves.
(i) Channel migration.
(j) High groundwater levels.

The phenomena are described in detail in SSG-18 [7], together with a methodology to derive the design basis conditions.

5.2. Scenarios that induce one or more of the following effects should be considered, as should the duration of the flood event:

(a) Wind waves and run-up effects.
(b) Hydrodynamic and other loading effects:
    (i) Hydrostatic load;
    (ii) Hydrodynamic load;
    (iii) Wave load;
    (iv) Buoyancy load (vertical hydrostatic load);
    (v) Debris load;
    (vi) Sediment load.
(c) Erosion and sedimentation.
(d) Concurrent site conditions, including adverse weather conditions.
(e) Groundwater ingress:
    (i) Seepage and groundwater inflow;
    (ii) Leakage.

5.3. The design should consider the potential damage to SSCs important to safety due to the infiltration of water into internal areas of the installation as well as the resulting water pressure on walls and foundations that could challenge their structural capacity or stability. Groundwater may affect the stability of soil or backfill. Deficiencies or blockages in site drainage systems also could cause enhanced flooding of the site.

5.4. The design should consider the dynamic and static effects of water, which can be damaging to the structures and foundations of a nuclear installation, as well as to the many systems and components located on the site. Moreover, there may be erosion at the site boundaries, scouring around structures or internal erosion of backfill due to the effects of groundwater.

Parameters characterizing the hazard

5.5. The storm surge analysis should include estimates of static water elevation, or a distribution of water elevation with a corresponding annual frequency of exceedance, depending on whether a deterministic or probabilistic method is used.

5.6. The wind wave analysis should include estimates of the increases in water level due to wind wave activity and wave run-up height along the beach or structures. In addition, relevant parameters (i.e. wave kinematics) associated with the dynamic effects of waves on structures should be considered. Loading and unloading analyses should include hydrodynamic effects, static loading effects, erosion and sedimentation, and other associated effects.

5.7. The tsunami flooding analysis should include estimates of the maximum water level, event duration, run-up height, inundation horizontal flood, backwater effects, minimum water level, and duration of the drawdown below the intake. Loading and unloading analyses should include hydrodynamic effects, static loading effects, water-borne missiles, erosion and sedimentation, and other associated effects.

5.8. For tsunamis induced by earthquakes in the vicinity of the site, uplift and subsidence of the Earth's surface should be taken into consideration in assessing potential negative impacts on the estimation of the water height in areas close to large earthquake rupture zones.

5.9. The seiche hazard analysis should include estimates of the maximum and minimum run-up heights, the duration, the static loading effects, and the hydrodynamic effects listed in para. 5.2.

5.10. The design against river flooding should consider all types of flooding, including dam failures, the duration of floods and the existence of flood protection and navigation systems. With regard to an estuary, the design should consider combinations of high tides, wave effects, high wind-driven water levels and high water levels in the river.

5.11. The design relating to flooding due to local precipitation should consider site grading, site and buildings drainage, sheet flow, and flow on the site from off-site areas. The design parameters should include flow rate and duration, peak water level, the variation of water levels over time, and mean water velocity to estimate the hydrodynamic forces and potential sedimentation and erosion on the site.

5.12. The parameters used to characterize floods due to the sudden release of impounded water should include the following:

(a) The anticipated flow rates during the entire flood event;
(b) The peak water level at the site and the variation of the water surface elevation over time;
(c) The potential for water intakes to become blocked or damaged;
(d) The dynamic and static forces resulting from debris or ice.

The parameters used to characterize flooding due to dam failure should also include warning times.

5.13. The parameters describing bores and mechanically induced waves should include the maximum run-up height, the associated duration and the impact of the tidal fluctuation.

5.14. High groundwater levels in the close vicinity of the site are generally a consequence of another phenomenon, such as an increase of water level near a river or sea, intense precipitation, or failure of water control structures. Parameters such as extreme groundwater level and the associated pressure on structures should be characterized.

5.15. Flooding due to local precipitation applies to all sites. The occurrence of other flood phenomena depends on the site location (e.g. by a river or a lake, on the coast, in an estuary) (see paras 5.30–5.33).

5.16. The tidal range should be determined for sites located on the coast, in estuaries and in river areas affected by tides.

Design parameters for external floods

5.17. Design basis flood conditions should be derived on the basis of the recommendations provided in paras 6.4–6.16 of SSG-18 [7]. Such conditions may result from one extreme event or, more often, from a combination of events. They are expressed in terms of water level, water velocity, flow pattern and groundwater level, as well as relevant parameters associated with the events generating the flooding, as described in paras 5.5–5.16. The action of water on structures may be static or dynamic, or there may be a combination of effects. In many cases, the effects of ice and debris transported by the flood and the waves (or surge) are important factors in the evaluation of pressure on structures.

5.18. SSCs important to safety should be protected from damage due to flooding. The design basis for such SSCs should be determined from flooding effects at their locations. It should be taken into account that local factors (e.g. site layout and topography, site grading, neighbouring structures, flow directions, intake structures, configuration of the ultimate heat sink) may influence the loading conditions.

5.19. A drawdown of the sea level may result from a surge, a seiche or a tsunami. The effects associated with low water levels, including drawdown, on items important to safety (including the ultimate heat sink) should be considered.

5.20. In the event of extreme precipitation at the site, the drainage system is relied on, and the design should include an adequate margin. Deficiencies or blockages in site drainage systems should be considered in flooding analysis.

Means of protection against external floods

5.21. The nuclear installation should be protected against the design basis flood by one or more of the following means of protection:

(a) The 'dry site' concept (see paras 4.11–4.13), where the elevation of the installation, including all items important to safety, is above the design basis flood level with an adequate margin.
(b) Permanent barriers, such as flood walls, designed to prevent flood water from affecting SSCs important to safety.
(c) Other engineered features to protect SSCs important to safety that could be affected by flood, including the following:
    (i) Breakwaters;
    (ii) Site grading and drainage systems;
    (iii) Watertight doors and penetrations;
    (iv) Temporary watertight barriers, such as aqua dams, sandbags and inflatable berms, to be installed when necessary.

Permanent protective measures should be preferred over temporary protective measures.

5.22. For new nuclear installations, all SSCs ultimately necessary to prevent core damage, an early radioactive release or a large radioactive release should be located at an elevation higher than the design basis flood; alternatively, adequate engineered safety features (e.g. watertight doors) should be in place to protect these SSCs and ensure that mitigating actions can be maintained. For new nuclear installations, a dry site is preferred over a site protected by permanent external barriers. For existing nuclear installations, only the second option may be possible.

5.23. If the dry site concept cannot be applied to all SSCs important to safety, the layout should include permanent flood barriers with appropriate design bases and adequate margins (e.g. to protect against hydrodynamic effects, impacts from floating objects, seismic events).

5.24. Civil engineering structures (e.g. sea walls) that are permanent barriers for protecting SSCs important to safety against flooding should be designed to maintain their stability under the design basis loading conditions. The effects of flooding and other associated effects should be considered in assessing the potential failures of the structures.

5.25. Openings (e.g. watertight doors) that are permanent barriers should be designed to maintain their function under the design basis loading conditions.

5.26. External barriers and natural or artificial islands should be considered items important to safety and should be designed, constructed and maintained accordingly.

5.27. If any infill is necessary to raise the installation above the level of the design basis flood, this should be considered an item important to safety and should therefore be adequately designed and maintained.

5.28. A flood monitoring system should be provided that is able to detect conditions indicating the potential for flooding of the site. When feasible, the warning time should be sufficient to bring the installation to a safe condition together with the implementation of appropriate procedures. Specific operating procedures in response to the real-time monitoring data on the flooding conditions should be established.

5.29. The flood monitoring system should be designed to withstand the design basis flooding. If necessary, protection of the monitoring system from damage due to hydrodynamic forces and collisions of floating bodies should be provided.

Coastal sites

5.30. For a nuclear installation located at the coast, the following effects associated with design loading conditions should be considered:

(a) Run-up (sea water level);
(b) Drawdown;
(c) Hydrostatic forces, hydrodynamic forces and wave forces;
(d) Buoyancy;
(e) Collisions of floating bodies (e.g. logs, boats, barges);
(f) Erosion and sedimentation;
(g) Aftershock effects on flood protection equipment and mitigation equipment.

It should be taken into account that effects such as the movement of sand sediment and collisions of floating debris may occur simultaneously.

River sites

5.31. The design of a nuclear installation against river floods should consider similar loading phenomena, as appropriate, as for a coastal site (see para. 5.30) as well as the operational data relating to dams and the navigational system. The unique characteristics of river flooding include an extended duration of the flood event (weeks or months) and dam failure effects.

5.32. River floods in cold climates should be analysed for the formation of ice dams and the transport of large ice floes or sediment and debris that could physically damage structures, obstruct water intakes or damage the water drainage system. Potential ice dam formation and failure can flood the site or create low water conditions. Special consideration should be given to the potentially short warning times associated with flooding that might result from ice dam formation and failure.

Estuary sites

5.33. The design of a nuclear installation against estuary floods should consider similar loading phenomena, as appropriate, as for a coastal site or a river site (see paras 5.30–5.32). The unique characteristics of estuary flooding include a combination of the effects of river flooding and of coastal flooding, for example the combined effects of extreme high tides, wind wave, extreme precipitation and river flooding.

Assessment for beyond design basis external floods

5.34. Beyond design basis flooding should be defined by increasing the design basis flood level and considering the appropriate combination of events that may be associated with the flood.

5.35. For new nuclear installations, SSCs ultimately necessary to prevent an early radioactive release or a large radioactive release should be located at an elevation above the beyond design basis flood; alternatively, adequate engineered features should be in place to protect these SSCs and ensure that mitigating actions can be maintained. For existing nuclear installations, only the second option may be applicable.

EXTREME WINDS

Interface with the hazard evaluation

5.36. SSG-18 [7] provides recommendations on assessing extreme wind hazards due to strong ('straight line') winds, tropical storms (e.g. cyclones, typhoons, hurricanes), and tornadoes. For this Safety Guide, the output of interest from the assessment of wind hazards for the site evaluation is the hazard curves for wind speed (e.g. median, mean and fractiles, a discrete family of curves) in open terrain and at a specified height, usually 10 m above ground level.

5.37. The results of the hazard evaluation are used to determine the design basis wind speed. The values for the design basis wind speed should be consistent with the selected design basis external event policy of the regulatory body.[17]

5.38. Wind speeds should be averaged over time periods that are consistent with the natural frequencies of SSCs.[18] In addition, corrections for local topographical effects, if any, should be considered.

5.39. For some sites, in addition to design basis wind speeds corresponding to 'extreme' meteorological phenomena, those corresponding to 'rare'

---

[17] In some States, the design basis extreme wind speed is chosen on the basis of a 100 year return period (1% annual frequency of exceedance), whereas design rare events causing high winds (e.g. tornado, typhoon) are typically chosen with a much longer return period [7].

[18] For the structural design of nuclear installations, time averaging of gust speeds over 1–3 seconds is usually necessary.

meteorological phenomena, such as tornadoes and hurricanes [7], should also be considered. In design, the former type of phenomenon is usually considered an extreme condition and the latter a rare condition. The characteristics of 'extreme' and 'rare' meteorological phenomena are addressed in para. 2.9 of SSG-18 [7].

5.40. Unless there is clear evidence for a preferred direction of extreme winds, the wind at the design basis speed should normally be assumed to blow from the most unfavourable direction.

5.41. Beyond design basis wind speeds should be established at an appropriate annual frequency of exceedance that is less than that of the design basis external event.

Loading derivation

5.42. Structural loads derived from the wind speed and duration should be obtained in the form of the pressure or suction on surfaces exposed to the wind.

5.43. The actual wind forces should be determined from the wind velocity using shape factors. The vertical distribution of wind velocity should also be considered.

5.44. Wind loads can usually be treated as static loads for the structures that are normally built in nuclear installations. Dynamic structural effects are usually considered for structures whose natural frequencies are below 1 Hz.

5.45. The wind acting on the nuclear installation buildings is not the same as the wind in open terrain. Interference effects, such as sheltering by other buildings and Venturi effects in passages between buildings, can significantly affect the wind generated pressures. These effects can result in an increase in wind speed through a constricted space or a decrease in wind speed due to obstructions. The channelling of winds around structures may have an important influence on the wind forces. High winds have been known to cause the collapse of cooling towers as a consequence of a 'group effect' (i.e. due to interaction between building structures, even though they were individually designed to withstand an even higher wind speed). These effects should be considered in the design.

5.46. The combinations of wind induced loads with other design loads may vary depending on the origin of the wind. It is common practice to use larger load factors for 'straight line' extreme wind loads than for wind loads derived from rare meteorological phenomena such as hurricanes and tornadoes. In the case of rotational wind due to tornadoes, the direction of wind on one surface of a structure could be different from — or even opposite to — that on another surface. Such loading conditions should be considered in the design.

Design and qualification methods

Local response

5.47. The first set of failure modes that should be considered corresponds to local structural failures at the surfaces directly exposed to wind pressure or suction forces. These include portions of the building enclosure (e.g. walls, façade panels, roof panels, doors) used to transfer the wind loads to the building's main structural system. This type of local structural failure is the most commonly observed during extreme wind events. Typically, these failures do not cause a major collapse, but they might affect SSCs located in the immediate vicinity of the failure and, in addition, produce a change in the ambient pressures within the building. Wind capacity analysis for these failure modes should be performed, which usually involves assessment of the structural capacity of the enclosure elements themselves and assessment of the mechanical capacity of the connection to the main structural system.

5.48. In analysing the failure of SSCs within buildings, the design should conservatively assume that a failure in the enclosure causes the failure of all sensitive equipment intended to be protected by the portion of the enclosure that has failed.

Global response

5.49. The second set of failure modes that should be considered corresponds to the global failure or global instability of the main structural system of metal frame buildings under the wind loads. These failure modes could produce a major collapse of the building. Wind capacity analysis for global failure modes should assess the structural capacity of the main structural system under the wind loads.

Impact by wind-borne missiles

5.50. The aerodynamic forces produced by extreme winds can accelerate objects and produce missiles that impact SSCs. The resulting impact loads constitute one of the principal loading effects of extreme winds, and they should be considered in the design.

5.51. Wind-borne missile analysis should be performed to identify the potential missiles. Such an analysis usually follows a deterministic approach and considers a spectrum of different missile types and maximum velocities. Administrative procedures can reduce the spectrum of missile types to be considered; however, such procedures should be credited only if it can be ensured that they are continuously effective.

5.52. The effects of a missile impact include a local response (e.g. penetration[19], spalling[20], scabbing, perforation) and an overall ('global') response of the impacted SSC (e.g. dynamic shear effects at the edge supports of an impacted wall). Local response effects should be estimated, taking into account the missile type and target materials. The overall response, when relevant, should be estimated through dynamic analysis, taking into account deformation of the missile or the time history of the impact force.

5.53. The velocity and orientation of the missile are important input parameters to determine missile impact effects. In general, the missile impact should be assumed to have a velocity vector perpendicular to the target surface, and the missile axis should be assumed to be collinear with the velocity vector.

Atmospheric pressure changes

5.54. Loads from atmospheric pressure change result from the variation in atmospheric pressure as a vortex moves over a structure. Such loads should be considered, especially for tornadoes, which have a combination of relatively high translational storm speed and a significant pressure drop in the centre of a rapidly rotating vortex.

5.55. Loads from atmospheric pressure change should be estimated using a model of the tornado wind field and knowledge of the rate at which the structure is able to vent.

Dust storms and sandstorms

5.56. For the design of a nuclear installation against dust storms and sandstorms, in addition to the associated wind speeds, other parameters from the hazard

---

[19] Penetration is the state when an impacting missile has formed a notch on the impact face but has not perforated the target.

[20] Spalling is the ejection of target material from an impact face as a result of a missile impact.

evaluation are needed, such as the duration of the storm, the chemical and physical properties of the dust or sand particles, and the expected dust or sand loading (in mg/m^3) of the air during the storm.

5.57. The design against dust storms and sandstorms should take into account the following:

(a) The increase in the effective air density, which produces larger wind pressures on exposed surfaces;
(b) Effects due to the accumulation of dust or sand, which could increase loads on roofs and walls and block access routes;
(c) Potential clogging of filters in air intakes for heating, ventilation and air-conditioning systems or emergency diesel generators;
(d) Abrasive and corrosive effects on equipment, especially in the long term;
(e) Difficulties in performing radiation monitoring during dust storms and sandstorms;
(f) On-site management and communications under reduced visibility conditions;
(g) Sand deposition in the ultimate heat sink.

Miscellaneous effects from extreme winds

5.58. As well as affecting the structural integrity of SSCs, extreme winds can cause other effects that should be considered in the design of a nuclear installation. Examples are as follows:

(a) Pressure differentials could affect the ventilation system.
(b) Particles carried by the wind could damage exposed surfaces and prevent the functioning of components and equipment.
(c) Saltwater spray could jeopardize the functionality of electrical equipment.
(d) Electrically conductive missiles (e.g. panels of steel sheet) could cause short circuits at the switchyard.

5.59. The ultimate heat sink and associated transport systems should be evaluated to ensure that any changes in water level caused by extreme winds will not prevent the transport and absorption of residual heat. Credible combinations of effects should be considered, when appropriate.

5.60. The effects of wind on SSCs not important to safety that could cause interactions with structures important to safety could also be of concern, for example the collapse of large cranes located outside structures important to safety.

A dedicated analysis should be performed, and adequate mitigation methods, such as physical separation or protective structures, should be provided as necessary.

Means of protection against extreme winds

5.61. Building structures, as normally designed for nuclear installations, provide a good level of protection against wind hazards for items important to safety. Consequently, for design against extreme winds, items important to safety should be located inside such buildings, leaving as few as possible of these items exposed to the outside environment.

5.62. Sensitive items important to safety that are located outside buildings should be protected against wind-borne missiles. Sensitive items include instrumentation, small diameter piping and tubing, glass or ceramic pieces, dials and gauges, exposed belts, and chains or couplings on motors. The level of protection should be commensurate with the spectrum of missile types and maximum velocities considered in the design. Adequate immobilization of equipment or materials located outdoors to prevent wind-borne missiles should also be considered.

Assessment for beyond design basis extreme winds

5.63. Assessment for beyond design basis wind should be performed for SSCs used (a) for the confinement of radioactive material or (b) for mitigating the consequences of an accident caused by extreme winds or associated hazards.

5.64. The methods used in the assessment for beyond design basis wind should normally be the same as those used for design basis wind, although there will be differences in engineering approaches that apply realistic assumptions as well as in the acceptance criteria and the material properties used in the assessment (see Section 4).

OTHER EXTREME METEOROLOGICAL CONDITIONS

5.65. SSG-18 [7] gives guidance for a site specific review of extreme meteorological conditions, including the following:

(a) Extreme air temperature and humidity;
(b) Extreme water temperature;
(c) Snowpack;

(d) Freezing precipitation and frost related phenomena;

(e) Lightning.

Other hazards may be connected with these conditions, such as hail and frazil ice. Frazil ice can block intake screens, pumps, valves and control equipment [7] and can represent a hazard to the ultimate heat sink. In some cases, an estimate may also be necessary of the low flow rate and the low water level resulting from the most severe drought considered reasonably possible in the region. The potential causes of such conditions include water evaporation, rainfall deficit, obstruction of channels, downstream failure of water control structures, and anthropogenic effects such as the pumping of groundwater.

5.66. The temperature of a river may vary greatly during the different seasons and may be affected by extreme weather temperature if it occurs for a sufficient period of time (days or weeks). The design of a nuclear installation located next to a river should take into account the effects of changes in river water temperature (which follows the weather temperature with a relatively short delay) that might affect the installation. It should be taken into account that a high river temperature might itself trigger restrictions or protective measures (e.g. reactor shutdown, power reduction).

5.67. Damage due to the extreme meteorological conditions described in para. 5.65 is usually represented in the analysis by the unavailability of the electrical power grid or the emergency power supply. Hazards such as snow could also affect ventilation intakes and exhausts, structural loading, air intakes on diesel generators, access to safety related facilities, and the mobility of emergency vehicles. Extreme air temperature, water temperature or atmospheric moisture could affect the heating, ventilation and air-conditioning systems of rooms housing items important to safety (especially electronic equipment) and the availability of the ultimate heat sink. The potential for these effects should be considered in the design and safety analysis of the installation.

5.68. Damage caused by lightning can be very extensive; therefore, protection from lightning should be considered.

Loading derivation

5.69. Environmental parameters for extreme meteorological conditions should be obtained from the hazard evaluation. These parameters include the duration of such conditions, their periodicity and their reasonable combination with other meteorological conditions, such as wind or precipitation, and biological conditions.

Design methods and means of protection

5.70. Unless special national codes or standards are available for the design of nuclear installations in relation to the extreme meteorological conditions described in para. 5.65, the structural design should follow the codes and standards for conventional buildings. Equipment should be qualified in accordance with its safety classification and external event classification.

5.71. The effect of snow on ventilation intakes and exhausts, roof design, diesel generator air intakes, access to safety related facilities, and the mobility of emergency vehicles should be considered in the design and safety analysis of the installation.

5.72. The effect of extreme air temperatures and water temperatures on items important to safety, especially electronic equipment, and on the availability of the ultimate heat sink should be considered in the design and safety analysis of the installation.

5.73. Lightning can cause various failure modes depending on its properties (e.g. peak current, current rising time, time of half value, impulse charge, specific energy). Different types of lightning impulse (e.g. first positive, first negative, subsequent, long) are defined in lightning standards. The higher the peak current is, the easier it is caught. Therefore, a minimum peak current should also be defined to design the lightning protection. Thermal, mechanical, electrical and electromagnetic hazardous effects of different impulse types should be taken into consideration in the design. Special attention should be given to the electrical and electromagnetic effects of lightning, since these effects might affect the safety of the nuclear installation more than other effects.

5.74. Special protection from lightning should be designed and implemented, with periodic assessment of the dedicated protection means, in accordance with international industrial standards, special national codes and standards, or qualified modelling. The design of a nuclear installation should provide sufficient protection against both the conductive and radiative effects of lightning.

5.75. Intake structures for the heat transport systems directly associated with the ultimate heat sink should be designed to provide an adequate flow of cooling water during seasonal water level fluctuations, as well as under drought conditions.

5.76. The design of nuclear installations should take into account the effects of extreme weather conditions on make-up supplies, even when these supplies do not involve any extensive off-site capability. For example, effects such as the freezing of supply pipework should be considered, and trace heating should be provided, where appropriate.

5.77. Measures should be taken, through testing and/or analysis, to confirm that the facilities provided to transfer heat to the ultimate heat sink will retain their capability under extreme meteorological conditions, particularly if there are long periods when the facilities are not used. These measures could include, for example, design measures to facilitate the monitoring of spray nozzles or intake screens to check that they are not blocked by ice. To prevent service water blockage due to frazil ice, measures to prevent frazil ice formation (e.g. outlet water recirculation to intakes, bar screen heating) should be implemented and alternative path(s) for the cooling water intake should be provided. Provision should be made for adequate instrumentation and alarms and relevant procedures and training.

Assessment for beyond design basis conditions

5.78. Extreme meteorological conditions that are beyond the design basis of the nuclear installation should be considered, taking into account predictions of climate variability and climate change that might affect the design basis parameters already considered. The predicted implications of climate change should also be taken into account when considering other beyond design basis external hazards that might be directly or indirectly affected by meteorological conditions.

VOLCANISM

5.79. Recommendations relating to the evaluation of volcanic hazards are provided in SSG-21 [8]. Table 1 of SSG-21 [8] comprises a list of volcanic phenomena together with their potentially adverse characteristics for nuclear installations. A nuclear installation should be protected against all volcano related hazards that have been identified.

5.80. First, it should be reconfirmed that adequate measures are available for all the phenomena associated with volcanoes identified in the hazard evaluation. In general, phenomena such as pyroclastic flows, lava flows, opening of new vents and ground deformation (including debris avalanches) are considered to be exclusionary conditions during the site selection stage. If these phenomena have not been screened out during the hazard evaluation, criteria relating to any protective measures should be discussed with the regulatory body.

Design methods and means of protection

5.81. The design envelope of the nuclear installation for external hazards may provide sufficient protection against some of the effects of volcanic phenomena. This should be verified for each individual effect using a conservative approach to take uncertainties into account.

5.82. If the effects of volcanic phenomena are not bounded by the external hazard design envelope of the nuclear installation, then additional design features or site protective measures should be provided.

5.83. Tephra fallout can result in static physical loads as well as produce abrasive and corrosive particles in air and water. The additional gravity loads on horizontal surfaces should be appropriately combined with other vertical loads. Tephra may also cause disruption of safety related SSCs by entering into orifices such as exhausts and intakes similar to sand and dust storms. Appropriate measures should be taken against these effects.

5.84. As noted in para. 5.80, massive flows, such as lava flows, pyroclastic flows, lahars and debris avalanches, should normally be screened out in the site selection process (see table 1 of SSG-21 [8]). With regard to nuclear installations, there is no credible precedent for protective measures against these phenomena.

5.85. Volcano generated missiles generally affect a limited area around the volcano; the nuclear installation site should be selected to be outside this area. However, design bases should still be derived for missiles that have a low probability of reaching the site. The effects of these missiles should be compared with the effects of other missiles (e.g. wind-borne missiles), and impact and potential fire hazards should both be considered. Parameters for volcano generated missiles, which should be obtained from the hazard evaluation, include mass, terminal velocity and temperature.

5.86. If hazards relating to gases and aerosols from volcanic eruption have been identified and a design basis has been derived, then design features and procedural measures should be provided. Parameters that should be obtained from the hazard evaluation include the type of gas or aerosol, its physical and chemical properties, and its predicted concentrations where SSCs important to safety (including the control room) are located.

5.87. Volcano induced flooding should be considered in a coordinated manner with other external flood hazards (see paras 5.1–5.35). Floods induced by volcanic activity may affect both coastal and inland sites. Tsunamis and seiches should be considered for coastal sites; however, crater lake failures and glacial burst may affect any site, coastal or inland. Parameters that should be obtained from the hazard evaluation are similar to those for floods from other causes.

5.88. Volcanic earthquakes should be considered in the seismic hazard analysis for the nuclear installation. If volcano seismic hazards at the site are higher than those associated with other sources of seismic activity, the ground motion from volcanoes should be evaluated.

Design basis conditions and beyond design basis conditions

5.89. Non-exclusionary volcanic hazards should be treated as design basis external event loads. If any exclusionary volcanic hazards cannot be adequately screened out with sufficient margins, these should be treated in the framework of beyond design basis external events.

EXTERNAL FIRE

5.90. Fire that originates outside the site (e.g. from fuel storage, vehicles, pipelines or chemical plants; from natural vegetation) could affect the safety of the nuclear installation. Site accessibility during an external fire should be considered. A specific analysis for coastal sites should consider the potential for oil being spilled into the sea (e.g. by a stricken vessel or an extraction platform). If necessary, appropriate measures should be taken to establish a fire exclusion zone. Recommendations on identifying and evaluating external fire hazards from human induced events are provided in NS-G-3.1 [9].

5.91. The design of the nuclear installation should prevent smoke and heat from fires of external origin from impairing the fulfilment of safety functions and the stability of safety related structures at the site.

5.92. The ventilation system should be designed to prevent smoke and heat from affecting redundant divisions of safety systems and causing effects (including on the actions of operating personnel) that could impair the fulfilment of a safety function.

5.93. If the effects of an aircraft crash at or near a nuclear installation are being considered (see paras 5.161–5.192), a fire hazard analysis of such an event should also be performed. Fires and smoke could occur at several locations because of the spreading of aircraft fuel and combustible debris, and this should be considered in the analysis. Special equipment, such as foam generators and entrenching tools, as well as specially trained on-site and off-site firefighting personnel may be used to prevent such fires from penetrating structures containing items important to safety.

Loading derivation

5.94. The fire hazard evaluation should take into account the characteristics of the postulated fire, including the radiant energy, the flame area and flame shape, the view factor from the target, the speed of propagation and the duration. Secondary effects such as spreading of smoke and gases should also be considered. Ignition by lofted firebrands and damage to ventilation inlet filters should also be considered in the fire hazard evaluation.

5.95. The effects of an external fire originating from sources such as fuel storage, vehicles or natural vegetation should be combined with normal operating loads. Fires as a consequence of scenarios such as an aircraft crash should be considered in the same load combination and with the same design assumptions as for the initiating event itself.

Design methods

5.96. The vulnerability of structures to the thermal conditions arising from large external fires should be assessed against the inherent capacity of the structures to withstand such conditions. The assessment should be based on the capacity of a structure to absorb thermal loads without exceeding the appropriate structural design criteria. The capacity of concrete to resist fires is mainly estimated on the basis of the thickness, the composition of aggregates, the reinforcing steel cover and the limiting temperature at the interior surface. The limiting structural criteria are often the temperature at the location of the first reinforcing steel bar and the ablation of the surface exposed to the fire.

5.97. Reinforced concrete structures designed to carry impact loads resulting from an aircraft crash are generally strong enough to resist failures of structural elements in external fire scenarios.

5.98. The capacity of steel structures exposed to large fires is limited. Therefore, structures that are important to safety should not be constructed using steel as load bearing elements. If the fire resistance of steel structures relies on separation from external cladding or on an intumescent cooling, it should be verified that such fire protection measures are not adversely affected by secondary effects associated with fire scenarios (e.g. explosion pressure waves, missiles).

5.99. Criteria concerning the interior surface temperature and the air temperature in affected locations should also be considered in the assessment in order to protect items important to safety. These criteria are usually not exceeded if sufficient thickness is already provided to satisfy other considerations. Design penetration of all types should also be checked.

5.100. In cases where thick concrete walls or slabs are exposed to fire, a structural analysis should be carried out, taking into account the temperature gradient due to the fire, plus any additional operating loads under fire conditions (e.g. extinguishing water). A load factor of unity may be used in the ultimate load design for postulated fire loading conditions.

5.101. The selection of materials for a nuclear installation should take into account international codes and standards on fire hazards and the fire resistance of materials subjected to flame, heat and other phenomena.

Means of protection

5.102. Protection of the installation against fires that originate outside the site should be achieved by minimizing the probability of an external fire and by providing protective measures against external fires, where necessary. Measures should be taken to reduce the amount of combustible material and inflammable material in the vicinity of the site and near access routes to the site; alternatively, adequate fire protection barriers should be installed. Vegetation that could propagate a fire in close proximity to the installation should be removed. Other design measures, such as the physical separation and redundancy of safety systems, separate fire compartments or other fire barriers, and fire detection and fire extinguishing systems (e.g. sprinkler systems), should also be provided as appropriate.

5.103. If the inherent capacity of a structure does not provide sufficient protection against the effects of external fires, an additional barrier or physical separation should be provided. Additionally, heat resistant cladding or intumescent coatings could be used to provide further protection for structural elements. However, it should be verified that such features are not affected by secondary effects.

5.104. The ventilation system should be capable of being isolated from the outside air by means of dampers, with alternative measures being provided to accomplish the functions of the ventilation system. The air intake and exhaust of a ventilation system serving one safety system should be separated from the air inlet and exhaust serving other redundant safety systems to ensure that an external fire does not prevent the performance of a safety function.

5.105. The design of the nuclear installation should ensure an adequate supply of air to all diesel generators and other emergency power sources necessary to perform safety functions. This objective should be met by separating the air intakes by distance and segregating them from exhaust air outlets.

5.106. Safety related cables and instrumentation and control systems that are vulnerable to heat flux, smoke and dust should be qualified, or else protected against such hazards.

Assessment for beyond design basis external fire

5.107. Fires outside the nuclear installation building that have the potential to affect several safety related structures (e.g. caused by fuel spillage from a large airplane crash), including the containment, should be treated within the framework of a beyond design basis external event.

EXTERNAL EXPLOSIONS

5.108. The word 'explosion' is used in this Safety Guide to generally describe all events involving chemical reactions of solids, liquids, vapours or gases that could cause a substantial increase in pressure (and, possibly, fire or heat) in the surrounding space. Explosions of gas or vapour clouds can affect the entire installation. An analysis of the ability of the structures in a nuclear installation to resist the effects of a gas cloud explosion should be performed to assess the capacity of such structures to withstand the overpressure (i.e. direct and drag) loads. Other effects should also be considered, including fire, heat flux, smoke and heated gases, ground and other vibratory motions, and missiles resulting from the explosion.

5.109. In general, the following effects of explosions should be considered when analysing the response of the nuclear installation:

(a) Incident pressure and reflected pressure;
(b) Time dependence of overpressure and drag pressure;
(c) Blast generated missiles;

(d) Blast induced ground motion (mainly from detonation);

(e) Heat and fire.

5.110. If the installation has been designed to accommodate the effects of missiles generated from other external events, such as a hurricane, tornado or aircraft crash, the effects of missiles generated by an explosion might have already been taken into account. However, if particularly threatening missiles produced by explosions have been identified, they should be considered in the design of the installation. If missiles from an aircraft crash or natural phenomena are not included in the design basis, the potential effects of missiles should be considered in the evaluation of external explosion hazards.

Interface with the hazard evaluation

5.111. Explosions during the processing, handling, transport or storage of potentially explosive substances outside safety related buildings should be considered in the site hazard evaluation, in accordance with NS-G-3.1 [9]. The explosion hazard can arise from stationary or mobile sources. The result of the explosion hazard evaluation should include a list of potential explosion sources, the associated amount and nature of the explosive substance, the distance to the site, and the direction from the source to the site. The annual frequency of explosion for each source might also be needed.

5.112. Design basis parameters to protect the nuclear installation against unacceptable damage by pressure waves from detonations should be determined using one of the following methods:

(a) If there is a potential source of explosion in the vicinity of the installation that could produce a pressure wave external event, as described in table III of NS-G-3.1 [9], the propagation of the wave to the installation, including reflected waves, should be calculated and the resulting pressure wave and associated drag force should be included in the design basis for the installation.

(b) If there is already a design requirement to provide protection against other events (e.g. tornadoes), a threshold value should be calculated for the corresponding overpressure. This value allows the calculation of safe distances (i.e. stand-off distances) between the installation and any potential source.

Loading derivation

5.113. Unlike the detonation of explosives[21], liquid, vapour and gaseous explosive materials exhibit a considerable variation in their blast pressure output. An explosion of such materials is in many cases incomplete, and only a portion of the total mass of the explosive (i.e. the effective charge weight) should be considered in relation to the detonation process. A conservative estimate should be made for the portion of the total mass assumed to detonate.

5.114. The potential for flame acceleration and overpressure generation due to obstacles in gas clouds should be studied. These obstacles include equipment, piping and structures. There might also be a potential for flame acceleration due to trees and bushes.

5.115. Loads and heat effects derived from external explosions should be combined with normal operating loads only.

Detonation

5.116. The blast pressure loads from explosions should be determined using established engineering techniques (i.e. as mainly developed for the evaluation of hazards for chemical plants), such as TNT equivalent, multienergy methods, the Baker–Strehlow method and computational fluid dynamics. In the case of solid detonation, the TNT equivalent technique is the most widely used approach. In the case of a gas or vapour cloud, other approaches may be more appropriate to enable the elevation of the explosion and the reaction characteristics to be taken into account.

5.117. For the purposes of structural design or assessment, the variation of both the incident blast wave and dynamic wind pressures over time should be considered, since the response of a structure subjected to a blast load depends on the time history of the loading as well as the dynamic response characteristics of the structure.

---

[21] A detonation of explosives is characterized by a sharp rise in pressure that expands from the centre of the detonation as a pressure wave impulse at or above the speed of sound in the transmission media. It is followed by a much lower amplitude negative pressure impulse, which is usually ignored in the design, and it is accompanied by a dynamic wind caused by air behind the pressure wave moving in the direction of the wave.

Deflagration

5.118.   Deflagration[22] loads are not as well defined as detonation loads. Deflagration loads should be obtained using the same procedures as for detonation loads but should be conservatively based on an appropriate reduced mass of deflagrating material.

5.119.   Fire should be considered a secondary effect of deflagration, and the recommendations provided in paras 5.90–5.107 and SSG-64 [10] should be followed.

Design and qualification methods

Design for postulated explosion effects

5.120.   Protection against the effects of an external explosion can be ensured by designing structures to withstand the effects of detonations or deflagrations. The design should involve the following steps:

(a)   Characterizing the blast pressure and dynamic (wind) pressure acting on the structure, including any reflection due to orientation of the walls. The time history of the pressure is needed.
(b)   Determining forces acting on the external surfaces of the structure.
(c)   Determining the resistance of the structure to the pattern of forces, assuming elastic or elastic–plastic behaviour. The resistance depends on acceptance criteria, defined in terms of material strain limits and structural deformation limits. It is common that the overall resistance is governed by local failures (e.g. of exterior wall panels).
(d)   Computing the structural response to the forces determined in (b). This can be done using simplified models (e.g. single degree of freedom models) or complex models (e.g. non-linear finite element computations). In either case, even when using quasi-static computations, the dynamic nature of the

---

[22] A deflagration normally results in a slow increase in pressure at the wave front and, compared with a detonation, has a longer duration and a peak pressure that decreases relatively slowly with distance. These characteristics are also influenced by the weather conditions (e.g. temperature inversion) and the topography, which both need to be considered. A major difference between deflagrations and detonations is the heat or fire load on the target structure. In general, the heat or fire load from a detonation is not considered a part of the design basis for a target structure but is considered as such for a deflagration.

loads and the structural response should be considered. The effective loads on structures due to blast and associated dynamic wind loads are a function not only of the dynamic characteristics of the load but also of the dynamic response characteristics of the structure.

(e) Comparing the structural response with the structural resistance and modifying the design, if necessary. In performing this comparison, the structural resistance determined in (c) might need to be reduced to take into account the structural capacity necessary to sustain normal operational loads.

(f) Checking the ability of the main structural system to carry loads transferred from the exterior surfaces that directly receive the explosion loads, if the main structural system is not included in the model used to compute the structural response in (d).

(g) Checking the overturning and sliding stability of the structure.

5.121. The minimum parameters used to define the response of a particular structure should include the load buildup time and its peak value, as well as the damping and maximum level of ductility exhibited by the structure during the response.

5.122. In evaluating blast effects, a distinction should be made between the local and global response of buildings. The local response is associated with the response of external wall elements relative to their supporting members (e.g. girt, purlin, beam, column). The global response is typically associated with the primary load carrying system, which normally includes frames, beams, columns, diagonal bracing, shear walls and floor diaphragms.

5.123. External walls or roof elements directly exposed to explosion loads should be explicitly assessed on the basis of their local response.

5.124. For global structural elements that make up the primary load path for the structure, the peaks of loads are clipped by the elastic–plastic behaviour of the external elements directly exposed to the explosion. In such cases, simplified approaches to checking the ability of the primary load path to carry loads transferred from the exterior surfaces can normally be used, if justified.

5.125. Vibratory loads induced in building structures by the explosion should be evaluated and, if significant, the relevant response spectra should be calculated for the dynamic design of components and equipment, in accordance with their external event classification.

5.126. Direct and indirect effects of the explosion on ventilation systems should be assessed. Even if these systems are inside a structure, the analysis should verify that the ducts and any dampers in the systems are not damaged by the pressure wave to the extent that safety functions cannot be fulfilled.

Design for stand-off distance

5.127. Protection against the effects of an external explosion can also be ensured by a suitable stand-off distance between the explosion source and the target SSC. Safe distance studies should be performed in the site hazard assessment described in NS-G-3.1 [9]. At the design stage, once the layout of the installation and the size of structures is known, the safe distances should be verified using more accurate information.

5.128. When calculating the distances necessary to provide protection by means of separation, the attenuation of peak overpressure and heat as a function of distance from the explosion source should be taken into account. The data available for TNT can reasonably be used for other solid explosive substances by using the appropriate TNT equivalence. The adequacy of the protection afforded should be carefully evaluated for mobile sources on transport routes in the site vicinity. A sufficient number of plausible locations for the explosion should be postulated in accordance with NS-G-3.1 [9] to ensure that the most serious credible scenario has been analysed.

Means of protection

5.129. Shielding structures other than buildings should be considered as a means of protecting against blast wave loads and heat. Such structures are most useful for explosions generated by vessel ruptures or by detonations, in which case these structures should be designed to intercept missiles and provide protection against explosion overpressure. In such cases, the shielding structure should be close to the protected building to avoid pressure refraction behind the wall.

5.130. The protective measures that should be considered in design include adding supporting structural members to increase resistance and reduce unsupported spans; using strong backing walls for increased resistance; bolting walls to roofs, floors and intersecting walls to improve overall structural integrity; and replacing or reinforcing doors and windows with blast resistant elements. Automatic measures to protect against pressure waves should be considered in the design of air intakes important to safety, depending on the maximum overpressure of the intakes. Alternatively, it should be demonstrated that the incoming pressure wave will not lead to loss of the intended safety functions.

Assessment for beyond design basis explosions

5.131.   The methods used in the assessment of beyond design basis explosions should normally be the same as those used for design basis explosions. The differences should be reflected in engineering approaches that apply realistic assumptions, as well as in the acceptance criteria and the material properties used in the assessment (see Section 4).

5.132.   Beyond design basis explosions should be defined by increasing the amount of explosive substances and/or reducing the stand-off distances with respect to the values for design basis external events.

TOXIC, FLAMMABLE, CORROSIVE AND ASPHYXIANT CHEMICALS AND THEIR MIXTURES IN AIR AND LIQUIDS

5.133.   The release of toxic, flammable, corrosive and asphyxiant chemicals might affect the nuclear installation both externally and internally, damaging safety related systems and/or impairing the actions of operating personnel. Corrosive fluids could also affect outdoor areas, such as switchyards, and consideration should also be given to electrical and electronic equipment located outside buildings.

Interface with the hazard evaluation

5.134.   NS-G-3.1 [9] provides recommendations on evaluating hazards from the release of hazardous fluids at or near the installation. This hazard can originate from stationary or mobile sources. The result of the hazard evaluation should be a list of the potential sources of release of toxic, flammable, corrosive or asphyxiant chemicals, and the characteristics of such releases (e.g. the form of release, the location of release, the amount and nature of the hazardous substance released). If the hazard cannot be screened out on the basis of safe distance or probabilistic considerations, the outcome of the hazard evaluation should be used to characterize the releases to be included in the design basis of the nuclear installation.

Atmospheric dispersion of hazardous chemicals

5.135.   After characterizing the release to be used for design, the atmospheric dispersion of the released chemicals should be calculated by means of a model that allows for temporal and spatial variation in the release parameters and air concentrations.

5.136.   In most cases, Gaussian plume models for continuous releases, or 'puff' dispersion models with Gaussian concentration distribution within the plume for quasi-instantaneous and short term releases, are used. At a minimum, the model should take into account the longitudinal, lateral and vertical dispersion of the release. Complex computational fluid dynamics modelling may be considered appropriate for scenarios involving hilly terrain.

5.137.   The calculation of atmospheric dispersion should consider different scenarios linked to the time distribution of meteorological conditions at the site (i.e. wind speed, atmospheric stability, wind direction, precipitation, insolation, cloudiness). The goal should be to obtain dilution factors[23] between the release point and the relevant locations on the site, usually the air intakes of buildings.

5.138.   Toxic, flammable, corrosive or asphyxiant gases and vapour clouds may be heavier or lighter than air. In boil-offs and slow leaks, the effects of density on vertical diffusion should be considered and should be adequately supported by experimental data or numerical simulation. The density of heavier-than-air gases should not be considered when turbulence effects are more significant than buoyancy effects (e.g. when a release is the result of a burst, when the released material goes into the turbulent air near buildings). Special consideration should be given to heavy gas clouds formed by cold gas–air mixtures (e.g. liquid ammonia and air) that could travel far without being dispersed by atmospheric turbulence.

5.139.   Beyond design basis releases should be defined by increasing the amount of substance released and/or by reducing the distance between the point of release and the installation compared with the values for design basis releases.

---

[23] Dilution is usually expressed relative to the source of the release. For example, it can be expressed as the average gas or vapour concentration at a point, divided by the release rate at the source or divided by the concentration at the source.

Design and qualification methods

5.140. Once a toxic, flammable, corrosive, or asphyxiant gas or vapour cloud has been postulated, dispersion calculations should be carried out to estimate the gas concentrations as the cloud drifts or flows across the installation site.

5.141. Airflows during both normal and exceptional conditions should be considered in the design, together with the volumes of all rooms sharing one ventilation system and the volume of the ventilation system itself.

5.142. To simplify the calculations, it can be assumed that the gas or vapour concentration in the cloud remains constant during interactions with air intakes. Furthermore, it may be assumed that the gas concentration is the same in all rooms sharing one ventilation system. These assumptions are conservative when they relate to estimates of gas concentration but are not conservative when they relate to estimates of recirculation time or determinations of the amount of bottled air supply necessary; for these purposes, a more refined analysis should be carried out.

5.143. In some designs, the ambient air in certain rooms is isolated from potentially contaminated air after an accidental release. In such cases, the in-leakage rate to the isolated environment determines the time needed to reach hazardous gas or vapour concentration levels. The in-leakage rates considered in the calculations should be confirmed by testing the constructed system, functioning under the same conditions as assumed in the design bases.

5.144. When credit is given to the removal of hazardous chemicals by filtration, adsorption or other equivalent means, the technical basis for the removal capability should be included in the design documentation.

5.145. Once the concentrations of chemicals inside buildings have been determined, they should be compared with limits established in national regulations to assess the potential consequences to human health. Where appropriate, the estimated concentrations should also be compared with equipment specifications to assess potential effects on equipment performance.

Means of protection

5.146. The control room and its emergency ventilation system should have a low-leakage design.

5.147. Where there is a known source of toxic, flammable, corrosive or asphyxiant gas or vapour, appropriate detection systems at control room air intakes should be provided. If gas concentrations exceed the prescribed limits, protective actions should be initiated with due regard to quick-acting materials such as chlorine gas. These protective actions should include filtering the incoming air, temporarily preventing the ingress of air by use of recirculation air systems and, where necessary, using a self-contained breathing apparatus.

5.148. Some types of toxic, flammable, corrosive or asphyxiant gas or vapour, such as those that could be released along transport routes (i.e. roads, railways, seas, rivers), cannot be identified in advance. The provision of systems capable of detecting all types of hazardous gas or vapour is not practicable; however, where multiple types of gas or vapour could be a hazard, consideration should be given to providing detectors that are as versatile as practicable (e.g. capable of detecting groups of gases such as halogens or hydrocarbons) and that are also able to detect a decrease in oxygen levels.

5.149. For nuclear power plants, the supplementary control room, which is remote from the main control room and has a separate air supply, should be capable of shutting down and monitoring the reactor. The access route from the main control room to the supplementary control room should be protected to allow the movement of operating personnel; alternatively, arrangements should be made for personnel access via a control point at which a breathing apparatus is provided. If the supplementary control room is credited in the safety analysis, the air intakes for the supplementary control room should be separated by distance from the main control room air intakes. These intakes should be positioned at a high level if heavy gases or vapours have to be considered. The effectiveness of separation may depend on the ability to detect the presence of a toxic or asphyxiant gas in a timely manner. Thus, the means of protection should be specifically designed for each site.

5.150. For corrosive chemicals, it should be demonstrated that, even at the maximum possible rate of corrosion, the inspection intervals are such that safety systems could not be impaired to the extent that loss of a safety function could occur before the affected system can be repaired. Protection of systems may be achieved by different means, including the following:

(a) Preventing standing contact between corrosive agents and corrodible surfaces;
(b) Providing detectors for corrosive gases that activate closure valves;
(c) Using protective coatings;

(d) Providing additional wall thickness to allow a certain amount of corrosion;

(e) Reducing intervals between inspections;

(f) Using a combination of the above means.

Specific protective measures should be determined on a case by case basis. In some cases, it might be sufficient to keep the air temperature or humidity within specified limits, thus slowing down corrosion rates. The adequacy of such an approach should be demonstrated.

Assessment for beyond design basis conditions

5.151. Methods in the assessment for beyond design basis releases should normally be the same as in the design for design basis releases. The differences should be reflected in engineering approaches that apply realistic assumptions and in the acceptance criteria (see Section 4).

RADIOLOGICAL HAZARDS FROM OTHER ON-SITE AND COLLOCATED INSTALLATIONS

5.152. The release of radioactive gases, liquids and aerosols from adjacent operating nuclear units or storage installations, from vehicles transporting new or spent fuel, and from other on-site and off-site sources constitutes a potential external hazard. The release of radioactive substances could affect the nuclear installation by damaging safety related systems and/or impeding the actions of operating personnel.

Interface with the hazard evaluation

5.153. NS-G-3.1 [9] provides recommendations on evaluating the hazards from releases of radioactive substances from other installations. IAEA Safety Standards Series No. NS-G-3.2, Dispersion of Radioactive Material in Air and Water and Consideration of Population Distribution in Site Evaluation for Nuclear Power Plants [15], provides recommendations on dispersion of radioactive material in air and water. The recommendations in these Safety Guides should be followed when identifying the external radioactive releases to be considered in the design of the installation.

5.154. Beyond design basis releases should be defined by increasing the amount of radioactive substance released and/or by reducing the distance between the point of release and the installation compared with the values for design basis releases.

Design and qualification methods

5.155. The design of the nuclear installation should consider all potential external radiological hazards and should aim to ensure that the protection and safety of installation personnel complies with the relevant requirements of IAEA Safety Standards Series No. GSR Part 3, Radiation Protection and Safety of Radiation Sources: International Basic Safety Standards [16], and with national regulatory requirements. In addition, the nuclear installation should be designed in such a way that minimizes spreading of radioactive material that reaches the installation.

5.156. In the case of the release of radioactive material to the atmosphere, the potential concentration of radionuclides inside the installation should be calculated on the basis of the meteorological conditions and the air exchange rates within the installation to determine the time dependent air concentration, from which radiation doses can be calculated. The extension time and the interaction time of the gas or vapour cloud should be determined on an installation specific basis. Special attention should be paid to releases of radioactive material that could reach the air intakes for the control room and for other locations where personnel are present.

5.157. If there are scenarios in which radioactive material might enter the cooling water intake, the effect on the installation and the possible exposure of operating personnel should be considered. Special attention should be paid to systems that dissipate heat from the installation, as they could contribute to the spread of the released radioactive material.

Means of protection

5.158. For any radioactive external hazard to be considered in the design, two basic means of protection should be considered by the designer: shielding against external radiation exposure and the use of filters to protect against internal radiation exposure.

5.159. Recommendations on the protection of operating personnel against asphyxiant and toxic gases are discussed in paras 5.146–5.150. These recommendations should also be followed, as appropriate, as a means of protection against radioactive gases, vapours and aerosols released externally to the nuclear installation.

Assessment for beyond design basis conditions

5.160. Methods in the assessment for beyond design basis releases should normally be the same as in the design for design basis releases, although there will be differences in the acceptance criteria (see Section 4).

AIRCRAFT CRASH

5.161. NS-G-3.1 [9] provides recommendations on evaluating the hazard from an aircraft crash on the nuclear installation site. The result of the site hazard evaluation, which is based on a screening procedure to identify the potential hazards associated with an aircraft crash, should be expressed in terms of either specific parameters for the aircraft (e.g. type, mass, velocity, stiffness) or load–time functions (with associated impact areas).

5.162. In the design of the nuclear installation against an accidental aircraft crash, alternative paths (normally one train) may be used to ensure the satisfactory performance of safety functions. Iterations in the design of SSCs may be necessary before the final external event classification is determined. All SSCs classified as external event category 1 and external event category 2 should be designed or evaluated for the postulated aircraft crash event.

5.163. The postulated aircraft crash should be analysed to determine its potential effects on the installation and the steps necessary to ensure that the radiological consequences remain below acceptable limits. The following effects should be considered:

(a) Localized structural damage due to the impact of extremely stiff parts of the aircraft (e.g. engine, landing gear), including penetration, spalling, scabbing and perforation ('local effects');
(b) Global structural damage, including excessive deformations or displacements that prevent the structure from performing its intended safety functions ('global effects');
(c) The functional failure of SSCs due to induced vibrations in structural members and safety related equipment ('vibration effects');
(d) The effects of crash initiated fires and explosions on SSCs;
(e) Where relevant, the effects of fuel or extinguishing water entering buildings (e.g. through the ventilation system) on criticality safety.

Loading derivation

5.164.   The characteristics of the primary missile (i.e. the aircraft), the secondary missiles (e.g. engines) and the structure that receives the impacts should be defined. These characteristics include the following:

(a)   Missile type, velocity and impact angles;
(b)   Missile mass and stiffness;
(c)   Size and location of the impact area;
(d)   The load capacity and global ductility (or local strain limits) of the structure;
(e)   The secondary effects of an impact (e.g. secondary missiles, debris).

5.165.   The location of the impact area and the impact angle depend on the topology of the surrounding landscape, the neighbouring buildings and the type of aircraft.

5.166.   The modelling of the structure may be different in the local area and in the global area. The local area is the impact area and the surrounding area, in which the structure reacts non-linearly; in the global area, however, linear material behaviour can be applied. The applicability of the approaches used in structural modelling should be validated.

5.167.   The material properties assigned to structural steel, steel reinforcement and concrete should represent the realistic ductility of the materials (as determined by testing) and should include the strain rate effects and strength development of concrete over time.

Load–time function

5.168.   For the impact analysis of stiff or massive structures, an equivalent load–time function from the perpendicular impact of a defined, deformable missile on a rigid target should be derived using an analytical approach. A smoothing process should be applied to filter out, as far as possible, the unavoidable spurious noise from the numerical integration. Physical high frequency effects should not be excluded from the load function.

5.169.   Load–time functions can be used to consider a design basis external event. In such cases, the engineering design rules should comply with the relevant national or international codes and standards and with proven engineering practice.

Missile–target interaction

5.170. For an aircraft impact on flexible structures, the load might be heavily influenced by the dynamic interaction between the missile and the target, which can be determined using a coupled analysis (missile–target interaction) in which the aircraft type, mass, stiffness, velocity and impact angle (as a deformable missile) should be modelled. The type of aircraft, mass and velocity may be specified by the regulatory body.

5.171. Stiff components, such as engines and landing gear, should be included in the model referred to in para. 5.170. The impact load is defined by the initial velocity of the missiles.

5.172. The local area of a flexible target should be modelled using a sufficient number of concrete volume elements through the thickness. The non-linear material behaviour of the concrete — including different values in tension and compression, and different strain rates and failure criteria — should be defined. As far as possible, the material parameters should be validated using existing test data.

5.173. In the local area of a flexible target, reinforcing steel (which is subject to bending and shear stresses) should be modelled as beam elements connected to the concrete.

5.174. The model of the local area of the target should be capable of assessing the effects of failure modes of the concrete (from spalling to perforation) and of plasticity and damage of the steel.

5.175. Outside the local area (i.e. the global area), the model of the structure can be simplified in terms of the types of element, the element details and the material properties.

5.176. An alternative approach to assessing the effects of secondary missiles and debris relies on the application of empirical and semi-empirical analytical formulas mainly derived for rigid missiles. The ranges of shape, mass, stiffness and velocity for which the formulas were developed do not usually coincide with those of interest for an aircraft impact on a nuclear installation. Therefore, engineering judgement on the applicability of this type of approach is necessary.

Vibration effects

5.177.   In-structure response spectra should be calculated for all the main structural elements of buildings that contain safety related items.

5.178.   For the calculation of the building responses, appropriate damping modelling should be used, with care taken to avoid unreasonable values in the high frequency range. The analysis time should be long enough to ensure that any dominating vibrations of the structure after the impact are included.

5.179.   Spurious noise in high frequencies should, as far as possible, be filtered out of the numerical analysis of the time histories that describe the induced vibrations before taking into account the induced vibration in the design of the components.

5.180.   Passive mechanical structures can normally sustain high frequencies without damage, and in some States a high frequency cut-off in the in-structure response spectra is used. This approach is generally used where structural layouts are well defined and take into account high structural damping at high frequencies and the presence of structural discontinuities. In addition, such an approach should only be used if the calculated displacement is lower than the defined acceptability threshold and the vibration is propagated over a distance in the structure.

Fire effects

5.181.   The outer wall of the structure should be designed to resist an aircraft crash. Neither the aircraft nor parts of it should perforate the outer wall. The consequences that might result from the release of fuel carried by the crashing aircraft should be assessed using engineering experience. The following should be considered in this assessment:

(a)   The fire load, which should be directly related to the amount of fuel carried by the reference aircraft at the target (corresponding to the assumed scenario of the aircraft refuelling for the route from the starting airport to the destination, minus the fuel consumption from take-off and cruising) and the potential involvement of other flammable materials from inside the aircraft (e.g. hand baggage, luggage, payload, plastic sheeting, seats, flammable materials in the aircraft structures), as well as other flammable materials present at the site;
(b)   External fireballs;
(c)   Pool fires;

(d) Entry of fuel into buildings important to safety through normal openings or, as vapour or aerosol, through air intake ducts, leading to subsequent fires;

(e) Entry of combustion products into distribution systems, thereby affecting personnel or causing malfunctions in the installation, such as electrical faults or failures in emergency diesel generators.

Miscellaneous aspects

5.182. When analysing the hazard from accidental aircraft crashes at a nuclear installation, the soil should be represented by a damped spring mass system. For normal foundations and site conditions, it is sufficient to consider the average dynamic soil conditions of the site, because the variation in soil properties is usually expected to have a negligible effect on such an analysis.

5.183. The masses of the structural members as well as the dead load of equipment should be considered in the numerical model. Fluid stored in tanks or pools can be represented as rigidly connected masses. Actual live loads should be considered, rather than the design live loads that are generally assumed.

5.184. Some energy is expended in crushing the impact area and the immediate surrounding area; consequently, damping in the global area should be lower than in other global dynamic load cases.

5.185. Structures that perform a confinement function in nuclear facilities should be designed to withstand (without perforation) the impact of an accidental aircraft crash.

5.186. Sensitivity studies should be performed to determine the range of consequences associated with an accidental aircraft crash at a nuclear installation and the most sensitive parameters. In addition, any computer codes used for non-linear analysis should be verified and validated.

Means of protection

5.187. When considering the protection of SSCs in a nuclear installation against an aircraft crash, the different local effects, global effects and vibration effects of the crash, as described in para. 5.163, should be taken into account. Vibration effects should be addressed by providing redundant and sufficiently separated components or by implementing vibration isolation measures.

5.188. Concrete structures that might receive a direct impact should be reinforced on both sides, with sufficient stirrups.

5.189. The reinforcement should be designed in accordance with the minimum and maximum values (e.g. compression, tension) of the potential internal forces, adequately combined with other prescribed load conditions.

5.190. Where local structural failure (including scabbing) could impair the performance of a safety function by causing damage to items important to safety, the following measures should be taken, either individually or in combination:

(a) The structural resistance of the structure (or its layout) should be improved by increasing the thickness and/or reinforcement (or, in the case of underground distribution systems, by increasing the earth covering), by adding missile shields or other obstacles, or by other appropriate measures.

(b) Redundant equipment should be located in a separate area with an adequate separation distance (physical separation).

(c) The potentially affected items should be qualified for short transient loads. The equipment qualification should cover all critical failure modes (in terms of stability, integrity and functionality) identified in the safety analysis.

5.191. When the structural analysis is performed, it is not necessary to combine all design loads with the aircraft crash load. Generally, it is sufficient for only those loads expected to be present for a significant duration (i.e. dead loads, actual live loads and normal operating loads for equipment) to be combined with the aircraft crash load.

Assessment for beyond design basis aircraft crash

5.192. If a beyond design basis aircraft crash involving fully fuelled commercial airplanes is considered, acceptance criteria should be chosen such that, at a minimum, items important to safety of the nuclear installation that are necessary to prevent large or early release remain functional. A coupled analysis should be performed for beyond design basis aircraft crashes. Load–time functions can also be used to consider a beyond design basis aircraft crash. In both cases, a best estimate approach can be used for the margin assessment.

ELECTROMAGNETIC INTERFERENCE

5.193. Hazards relating to electromagnetic interference are described in paras 8.13–8.15 of NS-G-3.1 [9]. The protection of items important to safety in a nuclear installation against such hazards should be achieved through design or, where this is impracticable, through administrative measures such as the establishment of exclusion areas.

5.194. A clear distinction should be made between sources of electromagnetic interference that are off the site and those that originate within the installation. The design approaches and the administrative controls may be different depending on the location of the source.

5.195. The greater use of digital equipment in instrumentation and control systems in nuclear installations tends to increase vulnerability to electromagnetic interference. In addition, there continues to be a rapid increase in potential sources of electromagnetic interference. Therefore, the protection against electromagnetic interference provided in the nuclear installation should be reviewed more frequently than the protection provided for other types of hazard.

5.196. If potential sources of electromagnetic pulses have been identified as off-site hazards, the pathways followed by these pulses (e.g. through radiation or conduction) should be identified and protection should be provided accordingly. If an electromagnetic pulse source is of malevolent origin, the design should involve close cooperation with nuclear security specialists. Whatever the case, the aim should be to protect the nuclear installation against electromagnetic pulses of any origin using a single comprehensive design.

5.197. In designing shielding against electromagnetic interference, appropriate consideration should be given to material characteristics, surface finish, corrosion protection, galvanic compatibility and environmental protection.

5.198. Sources of electromagnetic interference may be stationary or mobile. Tests should be performed to verify the adequacy of design measures to protect SSCs against all such sources of electromagnetic interference. SSCs that are exposed to electromagnetic interference should be qualified by type testing.

5.199. Where protection against electromagnetic interference through design is not practicable, administrative controls such as exclusion areas should be established, and procedures should be developed for enforcing these measures.

BIOLOGICAL PHENOMENA

5.200.   Biological phenomena mainly affect the availability of cooling water from the ultimate heat sink and from the service water system as a consequence of clogging due to excessive growth of organisms such as algae, mussels, clams, fish and jellyfish. With regard to mussels and clams, their growth inside the seawater systems as well as their entry into these systems from outside should be considered. Malfunctions in ventilation systems have also occurred because of clogging by leaves or insects in the filters. In some cases, instrumentation and control cables have been attacked by rats or bacteria. Corrosion effects and accelerated ageing of steel structures exposed to the marine environment can be induced by sulphate-reducing bacteria. IAEA Safety Standards Series No. SSG-56, Design of the Reactor Coolant System and Associated Systems for Nuclear Power Plants [17], provides recommendations on how to deal with such hazards in the design of specific items important to safety.

5.201.   Scenarios affecting the availability of cooling water have usually been combined with flooding that causes the sudden relocation of marine growth deposited in other areas, which then leads to clogging of the water intake. Strong winds can cause the clogging of air intakes by leaves or insects.

Design methods and means of protection

5.202.   The first step in the evaluation of hazards due to biological phenomena should be an analysis of the environmental conditions. A monitoring regime should be established that takes account of the rate of growth of the biological matter and of the need for control measures (passive or active).

5.203.   Specific design provisions should be made to prevent the clogging of air intakes and water intakes, as appropriate. Suitable screens should be provided at air intakes and water intakes, or redundant paths for the intake of clean air and cooling water should be provided.

5.204.   Measures should also be taken to prevent vegetation and other organisms from entering cooling systems. Major blockages might occur as a result of rare accumulations of vegetation or seaweed loosened by a storm, by shoals of fish (which can rapidly block the screening systems), or by flotsam. The intake structure should be designed to prevent marine organisms from approaching close enough to be caught in the suction flow and trapped against the intake screens.

5.205.   Fixed screens may be provided in the intake channels or at the pump house to prevent the ingress of large fish or clumps of seaweed. Outer screens should be designed with sufficient strength to prevent large debris, mammals, fish, and alligators and other reptiles from entering the cooling water system. In addition, a second screening stage should be considered, using measures such as rotating drum screens. A third stage of filtration using fine strainers is also likely to be needed, depending on the service water characteristics and the design of the heat exchanger.

5.206.   Despite the measures described in para. 5.205, a total blockage might still be possible. If the external event affects a considerable proportion of the site or the shoreline, even alternative intakes might not provide sufficient protection against blockages. For such cases, an alternative ultimate heat sink or diverse water intakes should be provided.

5.207.   Cooling water used in condensers and in heat transport systems directly associated with the ultimate heat sink should be adequately treated to inhibit the growth of organisms within cooling circuits. Further design features should be provided to facilitate the cleaning of air intakes and water intakes.

5.208.   Provision should be made for frequent biological monitoring of the ultimate heat sink to give an early warning of changes that might significantly affect its performance. For example, the introduction of new strains of seaweed with different growth habits or greater tolerance to cooling water conditions can affect the availability of water.

5.209.   Dedicated operating and maintenance procedures should be developed for monitoring biological phenomena that might affect the safety of the nuclear installation and for preventing and mitigating accidents that might be caused by such phenomena. Control measures include treatment of biological phenomena using biocides or the use of sacrificial (i.e. replaceable) systems.

HAZARDS ASSOCIATED WITH FLOATING BODIES AND
HAZARDOUS LIQUIDS

5.210.   The design of the ultimate heat sink and the water intake for the service water systems important to safety should take into account that some components may be outside the site boundary and, in some cases, spread over a wide area.

5.211.   The collision of floating bodies with water intakes or with structures of the ultimate heat sink either is the result of specific scenarios (e.g. a ship collision) or is associated with more complex external event scenarios (e.g. ice and logs during a flood), as described in SSG-18 [7] and NS-G-3.1 [9]. Loads from colliding ships or the impact of debris ice should be combined with other loads depending on the originating scenario (mainly flooding) and the dependencies between these events. In addition to the actual collision event, associated phenomena should also be considered, such as oil spills or releases of corrosive fluids from ships, which could affect the availability or quality of cooling water. Hazardous fluids or particles can be released by ship collision or leakages of pipelines or offshore installations.

Interface with the hazard evaluation

5.212.   NS-G-3.1 [9] provides guidance on the evaluation of hazards from ship collisions and defines the important parameters that should be considered in the design basis, if the hazard is relevant for a site. When damage due to direct impact cannot be avoided by the implementation of preventive and protective measures, a vessel impact design basis should be established that is based on the present and predicted future traffic in the waterway. This design basis is normally specified in terms of a vessel size and an impact velocity.

Loading derivation

5.213.   For design purposes, head-on bow collisions should be considered. Forces from sideways collisions are assumed to be enveloped by bow collision forces. Global collision loads should be in the direction of vessel travel. The impact force should be applied at the water level.

5.214.   For sites with a safety related intake of water from navigable water bodies, the effects of shipping accidents on the capability to fulfil the heat removal safety function should be considered [7]. Of primary concern is the potential for blockage of the intakes of the heat transport system that are directly associated with the ultimate heat sink, which might be caused by the sinking or grounding of ships or barges and the resulting obstruction of intake structure bays, canals or pipes that provide a conduit for water to the intake.

Design and qualification methods

5.215. The design of water intakes against ship collision and oil spills or releases of corrosive fluids or particles should be capable of providing an adequate level of performance under various environmental conditions.

5.216. For debris and ice, the dynamic action derived from the analysis of potential events should be applied to the structures that are intended to guarantee structural integrity.

5.217. For coastal sites, adequate protective measures should be designed in accordance with the codes and standards developed for traditional mooring and ship protective structures.

Means of protection

Preventive measures

5.218. Preventive measures against ship collision should be established in close cooperation with the navigation authorities. Prevention is achieved by providing assistance to navigation through the installation of navigational aids, the introduction of navigation regulations and/or the implementation of vessel traffic management systems. The probability of a collision of large vessels in normal cruising can be significantly reduced by the implementation of such measures.

5.219. Where possible, the loss of fundamental safety functions should be prevented by a water intake layout that gives due consideration to the concepts of diversity, redundancy and separation by distance.

Protective measures

5.220. Structures exposed to potential impacts should be designed to withstand the impact loads; alternatively, a protective structure (e.g. a fender) should be provided to redirect the impact or to reduce the impact loads to levels that are not sufficient to cause damage.

5.221. If the resistance of the structure or the protection system is higher than the vessel crushing force, the vessel will crush and the impact energy will be primarily dissipated by deformation of the vessel. This could result in spillage of fuel oil or other chemicals. Therefore, the design of any protection system should consider not only the protection of the structure but also the preservation, to the maximum extent possible, of the vessel to avoid spillage or blockage of the water intake.

5.222. Several types of protective structure are commonly used in ports or waterways. Many of them can be adapted to protect water intakes and components of the ultimate heat sink (e.g. fender systems, pile supported systems, dolphin protection or floating protection systems). Similar systems should also be developed to prevent direct debris impact or buildup of ice.

5.223. Where a potential direct collision with the intake structure is of concern, measures should be taken to maintain the supply of cooling water and ensure the capability of the ultimate heat sink. The effects of the collision (e.g. induced vibration during impact) on components of the heat transport systems directly associated with the ultimate heat sink should also be considered.

Mitigation measures

5.224. Adequate measures should be taken to mitigate the effects of the potential spillage of liquids that could readily mix with the intake water and result in damage to the heat transport system or could seriously degrade the heat transfer capability. For oil spills, protection should be provided by the proper submergence of pump intake parts. However, in cases involving shallow submergence, special measures such as booms or skimmers that keep the oil at a safe distance from the pump intake parts should be implemented. Such measures may also be necessary if the potential for ignition of the oil or other fluid is of concern.

5.225. If the blockage of an intake is possible to the extent that the minimum flow necessary for heat removal cannot be ensured, then either redundant means of access to the ultimate heat sink or diverse means of fulfilling the design objective for the ultimate heat sink should be provided.

5.226. In the case of a significant hazard from ice, the static and dynamic action on the intakes should be considered. In addition, measures should be implemented to prevent ice accumulation within the intake structure.[24] Alternatively, a different method of providing cooling to the installation should be provided, for example from a different water source or by a closed loop, air cooled system.

---

[24] For example, in some States, when ice clogs the intake screens, warm cooling water is pumped from a discharge basin.

Assessment for beyond design basis conditions

5.227. Methods in the assessment for beyond design basis collisions should normally be the same as in the design for the design basis collision. The differences should be reflected in engineering approaches that apply realistic assumptions, as well as in the acceptance criteria and the material properties used in the assessment (see Section 4).

5.228. Beyond design basis external events should be defined by increasing the size of the floating body and/or the impact velocity with respect to the design basis values. The approach should account for potential changes, during the installation lifetime, in the physical limits that could impact the characteristics of the floating bodies (e.g. effects of changes in bathymetry due to, for example, sediment transport or climate change effects like changes in sea level).

OTHER EXTERNAL HAZARDS

5.229. Geotechnical hazards not associated with seismic loads should be considered in the design of the nuclear installation. In general, hazards such as subsidence or cavity collapse involve both soil improvement and foundation design; therefore the geotechnical hazard evaluation should be taken into account. Further recommendations are provided in IAEA Safety Standards Series No. NS-G-3.6, Geotechnical Aspects of Site Evaluation and Foundations for Nuclear Power Plants [18].

5.230. For hazards for which a specific beyond design basis external event has not been defined, a combination of hazards may be used for the assessment of beyond design basis external events.

COMBINATION OF HAZARDS

5.231. In general, external hazards should not be combined with other extreme loads unless one of the following conditions is present:

(a)   The external event triggers the occurrence of another external event, such as a tsunami being triggered by an earthquake or a submarine landslide. In this case, the effects of both external events on the nuclear installation should be considered with due regard to the time difference between the effects of these events at the site. This also includes multiple dependent events occurring concurrently (e.g. storm surge accompanying heavy rainfall, dam failures induced by heavy rainfall, serial upstream dam failures occurring in a cascading manner).

(b) The external event comprises several potential hazards that could all occur at the site. For example, a large airplane crash has the potential to cause impact, vibration, explosion and fire at the site, all of which should be considered.

(c) The external event causes a change from normal operation to accident conditions. This possibility should be evaluated and considered in the design of the nuclear installation.

(d) There are external hazards that are likely to occur at the same time (e.g. extreme cold and extreme snow; extreme wind, lightning and extreme precipitation).

6. SAFETY DESIGN PROVISIONS FOR NUCLEAR INSTALLATIONS OTHER THAN NUCLEAR POWER PLANTS

6.1. This Safety Guide addresses a broad range of nuclear installations, as described in para. 1.12. Although the requirements for research reactors and for nuclear fuel cycle facilities are described in Section 2, the main focus of this Safety Guide has been on nuclear power plants. The methodologies recommended for nuclear power plants are applicable to other nuclear installations by means of a graded approach.

6.2. A graded approach means that designs for external events (and evaluations for beyond design basis external events) are customized for different types of nuclear installation so that they are commensurate with the severity of the potential radiological consequences of the failure of the installation. A graded approach is used to provide higher levels of protection against events that could result in higher risk. States should decide what level of risk is acceptable and what level of protection against the external event should be provided.

6.3. The recommended methodology for applying a graded approach is to start with attributes relating to nuclear power plants and, if possible, adjust these attributes to installations with lower radiological consequences. If no such adjustment is justified, the recommendations for nuclear power plants are applicable, as far as practicable, to other types of nuclear installation. Decisions relating to beyond design basis external events for nuclear installations other than nuclear power plants should be based, as appropriate, on Requirement 22 of SSR-3 [2] and Requirement 21 of SSR-4 [3].

6.4. The likelihood that an external event would give rise to radiological consequences will depend on the characteristics of the nuclear installation (e.g. its use, design, construction, operation and layout) and on the external event itself. Such characteristics include the following:

(a) The amount, type (e.g. solid, liquid, gas) and status (e.g. stored) of the radioactive inventory at the site;
(b) The intrinsic hazard (e.g. criticality) associated with the physical processes and chemical processes (e.g. for fuel processing purposes) that take place at the installation;
(c) The thermal power of the nuclear installation, if applicable;
(d) The configuration of the installation for different kinds of activity;
(e) The distribution of radioactive sources in the installation (e.g. in research reactors, most of the radioactive inventory will be in the reactor core and fuel storage pool, while in processing and storage facilities it may be distributed throughout the facility);
(f) The changing nature of the configuration and layout of installations designed for experiments;
(g) The characteristics of engineered safety features for the prevention of accidents and for mitigation of the consequences of accidents (e.g. the containment, containment systems), and the need for active safety systems and/or operator actions for the prevention of accidents and for mitigation of the consequences of accidents;
(h) The characteristics of the structures of the nuclear installation and the means of confinement of radioactive material;
(i) The characteristics of the process or of the engineering features that might show a cliff edge effect in the event of an accident;
(j) Any characteristics of the site that are relevant in terms of the consequences of the dispersion of radioactive material to the atmosphere and the hydrosphere (e.g. size of the site, local population distribution);
(k) The potential for on-site and off-site contamination.

Depending on the criteria established by the regulatory body, some or all the above factors should be considered. In addition, the potential for fuel damage, the size and nature of radioactive releases and the resulting radiation doses may be the main factors of interest.

6.5. Prior to categorizing a nuclear installation (see para. 6.9), a conservative screening process should be applied, in which it is assumed that the complete radioactive inventory is released in an accident initiated by an external event. If the result of such a release is that there are no unacceptable radiological consequences[25] for workers, the public or the environment, and no specific regulatory requirements are applicable, the installation may be screened out from further consideration with respect to external events. In such cases, the design, construction, operation, maintenance and future review of the installation are subject to the State's codes and standards for commercial or industrial facilities.

6.6. If the results of the conservative screening process described in para. 6.5 show that the consequences of the potential release of the complete radioactive inventory might be unacceptable, further screening may be implemented (i.e. screening by magnitude and distance, and screening based on the probability of occurrence (see para. 3.3)). If the results demonstrate that there are no unacceptable radiological consequences, this should be documented, and the external events may be eliminated from further consideration.

6.7. The application of a graded approach should be based on the following information:

(a) The safety analysis report for the installation, which should be the primary source of information;
(b) The results of a probabilistic safety assessment, if one has been performed;
(c) The characteristics listed in para. 6.4.

6.8. For an existing installation, the graded approach may have been applied at the design stage or later, for example at a periodic safety review. If so, the assumptions on which the application of the graded approach was based and the resulting categorization (see para. 6.9) should be reviewed and verified. The results may range from no radiological consequences (associated with conventional installations) to high radiological consequences associated with nuclear power plants.

---

[25] Unacceptable radiological consequences are doses to workers or the public that exceed acceptable limits established by the State.

6.9. As a result of the application of the graded approach, three or more categories of installation may be defined (depending on State practice), as follows:

(a) The lowest category includes those installations with the lowest radiological consequences, and these may be regarded as similar to conventional facilities, such as hospitals.
(b) The highest category includes installations for which the consequences of accidents initiated by external events are comparable with those from nuclear power plants.
(c) There are often one or more intermediate categories of installation between the lowest and highest categories.

6.10. With regard to the categorization described in para. 6.9, the following should be taken into account during the design and evaluation of external event hazards:

(a) For installations in the lowest category, the design and evaluation for external events may be based on national building codes and standards, as established for important facilities within the State. Beyond design basis external events may be considered in a simplified manner.
(b) For installations in the highest category, design and evaluation procedures should be implemented in the same manner as for nuclear power plants, including the identification and evaluation of beyond design basis external events and cliff edge effects.
(c) For installations categorized in the intermediate hazard category, the following cases may be applicable:
    (i) If the evaluation of external event hazards is performed using methodologies similar to those described in this Safety Guide for nuclear power plants, two approaches may be implemented to determine a lower loading condition than for nuclear power plants:
        — If the external event hazard is defined probabilistically, a higher annual frequency of exceedance may be selected for design of the installation and evaluation of the installation for beyond design basis external events with, where relevant, the approval of the regulatory body.
        — If the external event hazard is defined deterministically, a loading condition less than that for nuclear power plants may be selected for design in accordance with the precedent set in the State for other non-radiologically hazardous facilities with, where relevant, the approval of the regulatory body; similarly, beyond design basis external event loading conditions may be selected for assessing margin.

(ii)   If the database and the methods recommended in this Safety Guide are found to be excessively complex and time and effort consuming for the nuclear installation in question, simplified methods for the evaluation of external event hazards, based on a more restricted data set, can be used. In such cases, the input parameters finally adopted for designing these installations should be commensurate with the reduced database and the simplification of the methods, with account taken of the fact that both these factors may tend to increase uncertainties.

7. APPLICATION OF THE MANAGEMENT SYSTEM TO THE DESIGN OF A NUCLEAR INSTALLATION AGAINST EXTERNAL EVENTS

7.1.   The management system to be established, applied and maintained by the operating organization is also required to ensure the quality and the control of processes and activities performed at each stage of the design (see Requirement 10 of IAEA Safety Standards Series No. GSR Part 2, Leadership and Management for Safety [19]).

7.2.   The design processes for the development of the concept, detailed plans, supporting calculations and specifications for a nuclear installation and its SSCs should be established and applied in accordance with the recommendations provided in paras 5.84–5.140 of IAEA Safety Standards Series No. GS-G-3.5, The Management System for Nuclear Installations [20]. The design process includes the following activities:

(a)   Initiating the design and specifying the scope;
(b)   Specifying the design requirements;
(c)   Selecting the principal designer;
(d)   Establishing a work control system and planning design activities;
(e)   Specifying and controlling design inputs;
(f)   Reviewing and selecting design concepts;
(g)   Selecting design tools and computer software;
(h)   Conducting conceptual analyses;
(i)   Conducting detailed design and production of design documentation;
(j)   Conducting detailed safety analyses;

(k)   Defining any limiting conditions for safe operation (sometimes referred to as the 'safe operating envelope');

(l)   Verifying and validating the design;

(m)   Applying configuration management;

(n)   Managing the design and controlling design changes.

7.3. Design requirements, inputs, outputs, changes, control and records should all be established in the design processes. The design outputs include specifications, drawings, procedures and instructions, including any information necessary to install or implement the designed SSCs or other safety measures.

7.4. Design inputs, outputs and changes should be verified. Individuals or groups performing design verification should be qualified to perform the original design. Persons carrying out verification should not have participated in the development of the original design (but they may be from the same organization). The extent of verification should be based on the complexity of the nuclear installation, the associated hazards and the uniqueness of the design. Typical design verification methods include design review, carrying out calculations by an alternative method and qualification testing. Previously proven designs need not be subject to verification unless they are intended for different applications or the performance criteria are different. Design records — including the final design, calculations, analyses and computer programs, and sources of design input that support design output — are normally used as supporting evidence that the design has been properly accomplished [20]. Computer codes and models used in design should be verified and validated through quality assurance, benchmarking, testing or simulation prior to use, if they have not already been proven through previous use [20]. The documentation of this verification and validation should demonstrate that such codes and models are comprehensive, precise, traceable, complete, consistent, verifiable and modifiable [21].

7.5. Interfaces among all organizations involved in the design should be identified, coordinated and controlled. The control of interfaces includes the assignment of responsibilities among, and the establishment of procedures for use by, participating internal and external organizations [21].

REFERENCES

[1] INTERNATIONAL ATOMIC ENERGY AGENCY, Safety of Nuclear Power Plants: Design, IAEA Safety Standards Series No. SSR-2/1 (Rev. 1), IAEA, Vienna (2016).

[2] INTERNATIONAL ATOMIC ENERGY AGENCY, Safety of Research Reactors, IAEA Safety Standards Series No. SSR-3, IAEA, Vienna (2016).

[3] INTERNATIONAL ATOMIC ENERGY AGENCY, Safety of Nuclear Fuel Cycle Facilities, IAEA Safety Standards Series No. SSR-4, IAEA, Vienna (2017).

[4] INTERNATIONAL ATOMIC ENERGY AGENCY, Site Evaluation for Nuclear Installations, IAEA Safety Standards Series No. SSR-1, IAEA, Vienna (2019).

[5] INTERNATIONAL ATOMIC ENERGY AGENCY, IAEA Safety Glossary: Terminology Used in Nuclear Safety and Radiation Protection, 2018 Edition, IAEA, Vienna (2019).

[6] INTERNATIONAL ATOMIC ENERGY AGENCY, Seismic Hazards in Site Evaluation for Nuclear Installations, IAEA Safety Standards Series No. SSG-9 (Rev. 1), IAEA Vienna (in preparation).

[7] INTERNATIONAL ATOMIC ENERGY AGENCY, WORLD METEOROLOGICAL ORGANIZATION, Meteorological and Hydrological Hazards in Site Evaluation for Nuclear Installations, IAEA Safety Standards Series No. SSG-18, IAEA, Vienna (2011).

[8] INTERNATIONAL ATOMIC ENERGY AGENCY, Volcanic Hazards in Site Evaluation for Nuclear Installations, IAEA Safety Standards Series No. SSG-21, IAEA, Vienna (2012).

[9] INTERNATIONAL ATOMIC ENERGY AGENCY, External Human Induced Events in Site Evaluation for Nuclear Power Plants, IAEA Safety Standards Series No. NS-G-3.1, IAEA, Vienna (2002). (A revision of this publication is in preparation.)

[10] INTERNATIONAL ATOMIC ENERGY AGENCY, Protection against Internal Hazards in the Design of Nuclear Power Plants, IAEA Safety Standards Series No. SSG-64, IAEA, Vienna (2021).

[11] INTERNATIONAL ATOMIC ENERGY AGENCY, Seismic Design for Nuclear Installations, IAEA Safety Standards Series No. SSG-67, IAEA, Vienna (2021).

[12] INTERNATIONAL ATOMIC ENERGY AGENCY, Periodic Safety Review for Nuclear Power Plants, IAEA Safety Standards Series No. SSG-25, IAEA, Vienna (2013).

[13] INTERNATIONAL ATOMIC ENERGY AGENCY, Engineering Safety Aspects of the Protection of Nuclear Power Plants against Sabotage, IAEA Nuclear Security Series No. 4, IAEA, Vienna (2007).

[14] INTERNATIONAL ATOMIC ENERGY AGENCY, Evaluation of Seismic Safety for Existing Nuclear Installations, IAEA Safety Standards Series No. NS-G-2.13, IAEA, Vienna (2009). (A revision of this publication is in preparation.)

[15] INTERNATIONAL ATOMIC ENERGY AGENCY, Dispersion of Radioactive Material in Air and Water and Consideration of Population Distribution in Site Evaluation for Nuclear Power Plants, IAEA Safety Standards Series No. NS-G-3.2, IAEA, Vienna (2002).

[16] EUROPEAN COMMISSION, FOOD AND AGRICULTURE ORGANIZATION OF THE UNITED NATIONS, INTERNATIONAL ATOMIC ENERGY AGENCY, INTERNATIONAL LABOUR ORGANIZATION, OECD NUCLEAR ENERGY AGENCY, PAN AMERICAN HEALTH ORGANIZATION, UNITED NATIONS ENVIRONMENT PROGRAMME, WORLD HEALTH ORGANIZATION, Radiation Protection and Safety of Radiation Sources: International Basic Safety Standards, IAEA Safety Standards Series No. GSR Part 3, IAEA, Vienna (2014).

[17] INTERNATIONAL ATOMIC ENERGY AGENCY, Design of the Reactor Coolant System and Associated Systems for Nuclear Power Plants, IAEA Safety Standards Series No. SSG-56, IAEA, Vienna (2020).

[18] INTERNATIONAL ATOMIC ENERGY AGENCY, Geotechnical Aspects of Site Evaluation and Foundations for Nuclear Power Plants, IAEA Safety Standards Series No. NS-G-3.6, IAEA, Vienna (2004).

[19] INTERNATIONAL ATOMIC ENERGY AGENCY, Leadership and Management for Safety, IAEA Safety Standards Series No. GSR Part 2, IAEA, Vienna (2016).

[20] INTERNATIONAL ATOMIC ENERGY AGENCY, The Management System for Nuclear Installations, IAEA Safety Standards Series No. GS-G-3.5, IAEA, Vienna (2009).

[21] INTERNATIONAL ATOMIC ENERGY AGENCY, Design of Instrumentation and Control Systems for Nuclear Power Plants, IAEA Safety Standards Series No. SSG-39, IAEA, Vienna (2016).

CONTRIBUTORS TO DRAFTING AND REVIEW

Altinyollar, A.	International Atomic Energy Agency
Azuma, K.	Nuclear Regulation Authority, Japan
Beltrán, F.	Consultant, Spain
Campbell, A.	Nuclear Regulatory Commission, United States of America
Coman, O.	International Atomic Energy Agency
Contri, P.	International Atomic Energy Agency
Fukushima, Y.	International Atomic Energy Agency
Gürpınar, A.	Consultant, Turkey
Henkel, F.O.	Consultant, Germany
Johnson, J.J.	Consultant, United States of America
Mahmood, H.	International Atomic Energy Agency
Morita, S.	International Atomic Energy Agency
Sollogoub, P.	Consultant, France

IAEA
International Atomic Energy Agency

ORDERING LOCALLY

IAEA priced publications may be purchased from the sources listed below or from major local booksellers.

Orders for unpriced publications should be made directly to the IAEA. The contact details are given at the end of this list.

NORTH AMERICA

Bernan / Rowman & Littlefield
15250 NBN Way, Blue Ridge Summit, PA 17214, USA
Telephone: +1 800 462 6420 • Fax: +1 800 338 4550
Email: orders@rowman.com • Web site: www.rowman.com/bernan

REST OF WORLD

Please contact your preferred local supplier, or our lead distributor:

Eurospan Group
Gray's Inn House
127 Clerkenwell Road
London EC1R 5DB
United Kingdom

Trade orders and enquiries:
Telephone: +44 (0)176 760 4972 • Fax: +44 (0)176 760 1640
Email: eurospan@turpin-distribution.com

Individual orders:
www.eurospanbookstore.com/iaea

For further information:
Telephone: +44 (0)207 240 0856 • Fax: +44 (0)207 379 0609
Email: info@eurospangroup.com • Web site: www.eurospangroup.com

Orders for both priced and unpriced publications may be addressed directly to:

Marketing and Sales Unit
International Atomic Energy Agency
Vienna International Centre, PO Box 100, 1400 Vienna, Austria
Telephone: +43 1 2600 22529 or 22530 • Fax: +43 1 26007 22529
Email: sales.publications@iaea.org • Web site: www.iaea.org/publications

FUNDAMENTAL SAFETY PRINCIPLES
IAEA Safety Standards Series No. SF-1
STI/PUB/1273 (21 pp.; 2006)
ISBN 92–0–110706–4   Price: €25.00

GOVERNMENTAL, LEGAL AND REGULATORY FRAMEWORK
FOR SAFETY
IAEA Safety Standards Series No. GSR Part 1 (Rev. 1)
STI/PUB/1713 (42 pp.; 2016)
ISBN 978–92–0–108815–4   Price: €48.00

LEADERSHIP AND MANAGEMENT FOR SAFETY
IAEA Safety Standards Series No. GSR Part 2
STI/PUB/1750 (26 pp.; 2016)
ISBN 978–92–0–104516–4   Price: €30.00

RADIATION PROTECTION AND SAFETY OF RADIATION SOURCES:
INTERNATIONAL BASIC SAFETY STANDARDS
IAEA Safety Standards Series No. GSR Part 3
STI/PUB/1578 (436 pp.; 2014)
ISBN 978–92–0–135310–8   Price: €68.00

SAFETY ASSESSMENT FOR FACILITIES AND ACTIVITIES
IAEA Safety Standards Series No. GSR Part 4 (Rev. 1)
STI/PUB/1714 (38 pp.; 2016)
ISBN 978–92–0–109115–4   Price: €49.00

PREDISPOSAL MANAGEMENT OF RADIOACTIVE WASTE
IAEA Safety Standards Series No. GSR Part 5
STI/PUB/1368 (38 pp.; 2009)
ISBN 978–92–0–111508–9   Price: €45.00

DECOMMISSIONING OF FACILITIES
IAEA Safety Standards Series No. GSR Part 6
STI/PUB/1652 (23 pp.; 2014)
ISBN 978–92–0–102614–9   Price: €25.00

PREPAREDNESS AND RESPONSE FOR A NUCLEAR OR
RADIOLOGICAL EMERGENCY
IAEA Safety Standards Series No. GSR Part 7
STI/PUB/1708 (102 pp.; 2015)
ISBN 978–92–0–105715–0   Price: €45.00

REGULATIONS FOR THE SAFE TRANSPORT OF RADIOACTIVE
MATERIAL, 2018 EDITION
IAEA Safety Standards Series No. SSR-6 (Rev. 1)
STI/PUB/1798 (165 pp.; 2018)
ISBN 978–92–0–107917–6   Price: €49.00

Safety through international standards

9789201360212 90000

DESIGN OF NUCLEAR INSTALLATION

ISBN 13: 9789201360212

9 789201 360212

INTERNATIONAL ATOMIC ENERGY AGENCY
VIENNA